THE UFO ABDUCTION BOOK

THE UFO ABDUCTION BOOK

Extraordinary Extraterrestrial Encounters of the Terrifying Kind

BRAD STEIGER

Foreword by Whitley Strieber

MUFON
Mutual UFO Network
est. 1969

This edition first published in 2022 by MUFON, an imprint of
Red Wheel/Weiser, LLC
With offices at:
65 Parker Street, Suite 7
Newburyport, MA 01950
www.redwheelweiser.com

ISBN: 978-1-59003-307-4

Library of Congress Cataloging-in-Publication Data available upon
request.

Cover design by Kathryn Sky-Peck
Cover image by Bigstock.com
Interior by Maureen Forys, Happenstance Type-O-Rama
Typeset in Pyke Text and Oswald

Printed in the United States of America
IBI
10 9 8 7 6 5 4 3 2 1

CONTENTS

FOREWORD

I well remember when I first read Brad Steiger's *The UFO Abductors* and found a window opening for me into a world of which I was already a part but which was still very much in danger of drowning me.

In 1987, when I published *Communion*, I had the thought in the back of my mind—really an assumption, nothing more—that things would quiet down. I'd had my encounter. The ordeal was over. It was time to look back and try to understand, to heal, and to do what I could to bring perspective and whatever healing I could to others.

Brad's book, with its thoroughgoing analysis, made me wonder anew if it was actually over—if, in fact, our ideas about the experience being pretty much a one-time thing were even correct. His research suggested otherwise, and looking back on this text today, it almost seems like the work of a psychic or a prophet.

Brad understood the experience to be much more deeply embedded in our lives than I did. He did not see it as a passing phenomenon. On the contrary, his research told him that it had a history and that there seemed to be some very deep motives involving the very core of human life—our souls, our sexuality, our true past, and our possible future.

In the years leading up to 2000, the abduction phenomenon was reported less and less frequently. As of 2020, a classic abduction report is rare, but now thousands of close encounter witnesses appear to have complex relationships with the phenomenon. These relationships are so strange and, for the most part, so far from what we might think of as conventional contact that it is not possible to identify them as alien contact.

So what happened, and what is happening now?

As Brad points out, the abductions seem to have had a number of objectives—among them, gathering sexual material in the form of eggs and sperm. Many people were shown babies after being abducted. As I and my wife experienced this aspect of the phenomenon, I would advocate on a personal level that it is not a fantasy. I had sexual material removed from my body. We experienced the anguish of a disappearing pregnancy.

Despite all the testimony that has been amassed, there is no definite, final proof that these things happened. However, if one takes a step back and imagines what we might do if we had the capacity to move between the stars and if we found another planet inhabited by an intelligent life form, what happened to us begins to make sense.

If that life form was so different from us that we could not understand it, our first act would be to survey it and its activities. Was it intelligent by our standards? How did its societies work? What sort of history did it have? We would also be interested in its level of intelligence, its aggressive tendencies, its religions, its arts, its military and governmental structures, its biology, and whether or not we could ever come into coherent contact with it.

If we were either ethical or predatory, or part of both, we would not want to disturb its cultural integrity. If we were looking for a companion species, we would initially keep ourselves hidden until they were used to us and had reason to feel empowered in any relationship with us.

Back in May of 1977, two physicists, T. B. H. Kuiper and M. Morris, published a brief speculation in the magazine

Science to the effect that aliens who were capable of traversing the stars would almost certainly keep themselves hidden if they came here. The reason was that, if they revealed themselves, we would cease to innovate culturally and scientifically. We would become supplicants. And the only reason they could possibly want anything to do with us would be to see what new we could bring—they would be interested in observing our independent experience of the world and our own process of discovery.

On the other hand, if they had some predatory reason to be here, such as removing our sexual material and using it for purposes of their own, they would also have a motive for remaining hidden.

And if both were true, which the evidence suggests is the case, then their motive for secrecy would be very powerful.

How might they use our sexual material, though? Of course, there may be uses for it that we don't understand. I have personally had quite extensive contact with real, physical people who were only close to human and had abilities that I witnessed personally that are not human abilities—specifically to affect the minds of others from a distance.

Were these people the result of genetic experimentation? I cannot draw that conclusion. I can only say what I witnessed: three of them lived in the woods behind a cabin my wife Anne and I owned in upstate New York. They were elusive but spent enough time close to the house and in the house as to leave us no doubt as to their actual, physical presence. One of them was frequently seen by us and others.

We then lost our cabin due to financial reverses and moved to a small condo we owned in San Antonio, Texas. Within days, these people were also living in the condo complex; in fact, they moved into an empty apartment just behind ours. When I informed the owner of the condo of this, he had them evicted—whereupon they went from door to door trying to sell the rest of us his furniture!

Two of them were full-sized human males, the third was the size of an older child, but, when you got a close look, he had the face of a late-middle-age adult.

After they left the condo, I never saw them again, but I have always thought that at least the shorter one must have been the result of some very strange processes.

There are other possibilities, because, during the abduction period, which reached its highest level of activity in the 1970s, then tapered off into the mid '90s, a very deep genetic pool must have been created. In fact, it might well be deep enough to re-create our species.

They seem to have concentrated their efforts in the United States. If so, then this would have been because of the diversity of the population and its mobility, with large numbers of people entering and leaving rural areas all the time.

Of course, all of this must remain speculative at this time. However, the evidence, as we understand it now, does fit these scenarios.

Brad Steiger's book does an excellent—indeed, essential—job of setting the scene and providing a foundation of the facts as they were known at the time it was written. It is to be hoped that, in time, this provocative and disturbing reality will come into ever greater focus.

—Whitley Strieber

CHAPTER ONE

The UFO Abductors

Lois is a businesswoman who lives in Arizona. "I am a regular churchgoer," she confides. "I am well respected in the community."

But in November 1978, along with her friend Gina, something happened to Lois that defies any easy explanation.

Lois and Gina were driving from Denver to Phoenix that November. While they were negotiating a lonely stretch of highway, they saw a hitchhiker standing at the side of the road. Although Lois had never before picked up a hitchhiker under any circumstances, she found herself slowing her automobile and stopping.

Even at a distance, there seemed to be something decidedly different about this hitchhiker. He was dressed casually in a plaid flannel shirt and jeans. He was so clean-shaven that it appeared as though he had no beard at all. His hair was long and blond, and his eyes were a brilliant sky-blue.

With a peculiarly warm smile of welcome, he leaned forward toward the window that Gina had opened. "I'm so glad you've come," he said in a soft, almost musical voice. "We've been waiting for you."

Lois remembers that there was something strangely familiar about the hitchhiker. "As crazy as it might sound," she said, "it was as if I had known him all my life. Gina is usually kind of shy around strangers—especially males—but she was smiling from ear to ear, as though she were greeting an old friend."

Her memory becomes very sketchy after the hitchhiker's statement that "they" had been waiting for them. She doesn't remember driving her car off the highway. It was more as though something had picked up her Pontiac and "floated" it toward a large craft that was hovering over the desert. Lois does, however, recall the stranger continuing to calm their fears by speaking to the two women in soothing, loving tones.

"Gina and I were separated," Lois stated, nodding her head in self-confirmation, "and I didn't see her again for quite a while. The beautiful hitchhiker disappeared, and I was surrounded by smallish entities with large, staring eyes. Their mouths were straight lines that I never saw open, but I seemed to hear voices—perhaps by telepathy."

Lois's clothing was removed, and she was told that she must lie on a table so that she might be examined.

"I obeyed everything they said without question," Lois said. "It was as if I had been hypnotized or something. I just did whatever they told me."

Lois has endured a number of physical illnesses in her life, and she found the beings' bedside manner no cooler than that of the average Earth physician. "Nothing they did hurt that much," she shrugged. "They took some blood, some samples of hair and skin here and there. I was not married at that time and had no children. I had a sense that a lot of the tests had to do with my fertility . . . or lack of it."

Lois was given a gown to wear. "They still seemed to be examining my clothing, my purse, everything that I had with me. I had been really nervous when the hitchhiker brought us aboard the space vehicle, but he kept saying over and over that Gina and I had nothing to fear, that they would not hurt us."

Although she still had not seen any sign of her companion, Lois was ushered into a room with a soft couch where she might relax. There appeared to be a vast window in the room that looked out toward the night sky.

"It really seemed as though we were moving through space," Lois commented. "It was like we were traveling to another world."

Lois had no concept of how long she might have been in that room, staring spellbound at what seemed to be colors and lights swirling past the window.

"The next thing I remember is when an elderly bearded man in a robe came into the room," Lois told us. "He looked like a regular, normal earthman. He was a bit above medium height and well proportioned."

She remembers only snatches of their conversation:

"Why have you taken me on board this craft?"

"Because you are one of us."

"Who are you?"

"You will remember in the fullness of time."

"Who are the little guys with the big heads and large eyes?"

"You will remember in the fullness of time."

"Why did they examine me as if they were doctors?"

"To see how you are. To see if you are well."

"Why do you care about me?"

"Because you have the key."

"I have a key? What key do I have?"

"You will remember when the time is right."

The next clear memory that Lois has is of waking in her car along a lonely desert road, far from the main highway. Gina was once again beside her. They were both hungry and thirsty. They looked at each other but said nothing.

After the two confused women drove to the nearest town, they discovered to their amazement that *five days* had passed. They hurriedly made reassuring calls to the anxious friends and relatives who had been expecting them.

In the years that have passed since the UFO abduction, Lois and Gina have seldom exchanged memories of the incredible experience. "Actually," Lois explained, "Gina refuses to discuss it at all. She regards the whole thing as some kind of high spiritual encounter that should not be denigrated by analyzing it. She thinks that we were blessed and that we should let it go at that."

Lois, on the other hand, remains disturbed by the abduction and by the thoughts of what may have occurred during all that missing time. Then, too, shortly after her marriage, she began to experience poltergeist phenomena in her home.

"Doors would open and close of their own volition," she stated. "Telephones would ring at weird times of the night, and there would never be anyone there. Lights would blink on and off. My husband went nuts with all this happening. He thought he had married a witch."

Lois would have periodic dreams of the blond, blue-eyed hitchhiker at the side of the road. He would open his arms in welcome, smile, and tell her that she had the "key."

One night when her first child was about five years old, the child awoke screaming, calling in terror for her mother. When Lois ran into the child's bedroom and took her in her arms, the girl sobbed: "Mommy, Mommy, they're in the house, and they want to see you. They say that they've come for the key."

Lois swore to us that she had never told her young daughter about the hitchhiker, the UFO, the abduction experience—anything. "I had never even told my husband about the experience," she stated firmly. "And yet here was my daughter repeating the same phrase to me that I had been told directly. I had the 'key.' Somehow the beings had contacted my daughter in her dreams and managed to get her to pass on a message to me."

After this nightmarish contact, the physical manifestations in the home became so violent that the family was forced to move. As an additional tragic side effect, Lois's marriage ended.

Lois has told no one of her experience. "I have kept all this weirdness to myself. I have virtually suffered in complete silence for all these years," she said. "I have really undergone a hell because of that experience. I still have no idea what they meant when they said that I had the 'key.' The whole thing is really driving me crazy."

Ultimately Lois made the decision to contact my wife Sherry and me in an effort to gain more answers about her abduction experience through the use of hypnotic regression.

And there are signs that Lois's experiences are far from over. Just recently, her second daughter, who is now four years old, looked up from her toys to remark: "Mommy, it won't be long now before the hitchhiker comes back to see you."

ALIEN ENCOUNTERS

For over twenty years, I have participated in the hypnotic regression of dozens of men and women who claim to have been abducted for brief periods of time by crew members from UFOs. Many of these contactees claimed to have been given some kind of medical examination. In some instances, they had peculiar punctures and markings in their flesh.

In the greatest number of alien encounters, the UFO-nauts were described as standing about five feet tall and being dressed in one-piece, tight-fitting jumpsuits.

Their skin was gray, or grayish green, and hairless.

Their faces were dominated by large eyes, very often with snakelike, slit pupils.

They had no discernible lips, just straight lines for mouths.

They seldom were described as having noses, just little snubs if at all; but usually the witnesses saw only nostrils nearly flush against the smooth faces.

Sometimes a percipient mentioned pointed ears but on many occasions commented on the absence of noticeable ears on the large, round heads. And, repeatedly, witnesses described an insignia of a flying serpent on a shoulder patch, a badge, a medallion, or a helmet.

It has seemed to me for years now that the UFOnauts may be reptilian or amphibian humanoids, and I believe that they have been interacting with Earth for millions of years.

An archaeological enigma with which I dealt extensively in earlier works (*Mysteries of Time* and *Space/The Reality Game*) had to do with what appear to be humanoid footprints that are found widely scattered in the geological strata suggestive of 250 million years ago.

There are many possible solutions to the mystery, but consider these two:

1. The amphibians evolved into a humanoid species that eventually developed a culture that ran its course or was destroyed in an Atlantis-type catastrophe—just after they had begun exploring extraterrestrial frontiers. Certain UFOnauts, then, may be the descendants of the survivors of that amphibian culture returning from their space colony to monitor the present dominant species on the home planet.

2. Some of today's UFOnauts may, in fact, be a highly advanced reptilian humanoid culture from an extraterrestrial world, who evolved into the dominant species on their planet millions of years ago—and who have interacted with our world's evolution as explorers, genetic engineers, or observers.

THE POLICEMAN AND THE UFO CREW

The concept of amphibious aliens visiting Earth would have drawn a raucous horselaugh from Ashland, Nebraska, City Patrolman Herbert Schirmer before the morning of December 3, 1967. Shortly after midnight on the morning that changed his life, Schirmer was struck by a mysterious ray when his cruiser approached a grounded UFO just past the intersection of Highways 6 and 63. Later, during a hypnotic session, Schirmer recalled boarding the UFO and communicating with its crew: "He [the alien in command] is telling me the ship is operated by reversible electromagnetism, which has something to do with gravity.

"I'm standing in a room that is about twenty-six feet in area. The ceiling is about six feet high. The lighting here is red, and it is coming down from strips in the ceiling.

"They have been observing us for a long period of time, and they think that if they slowly, slowly put out reports and have their contacts state the truth, it will help them. He is explaining to a certain extent that they want to puzzle people. They know they are being seen too frequently, and they are trying to confuse the public's mind.

"There is some kind of program of breeding analysis. Some people have been picked up and changed so they have agents in our world. They are very smart about the brain and how to change it.

"They were from four and a half to five and a half feet tall. Their uniforms were silver gray, very shiny.

"They wore boots, gloves, and they had a thumb and four fingers. Their suits came up around their heads like a pilot's cap. On the right side of their helmet they had a small antenna, just above where the ear would be. I never did see any of their ears. Their chests were bigger than ours. They were built very wiry and muscular. They walked with a very straight posture.

"Their eyes are the one thing that I will never forget. The pupil went up and down, like a slit. When they looked at me, they stared straight in my eyes. They didn't blink. It was real uncomfortable. There was just a very thin eyebrow above the eye, slanting up just a bit. Their noses were flat. Their mouths looked more like a slit than a regular mouth."

IMPLANTS LEFT IN HUMAN SKULLS

In 1969, I and my research associates, hypnotist Loring G. Williams and Glenn McWane, were bombarded with the claims of dozens of contactees who said that they had had an implant left somewhere in their skulls, usually just behind the left ear. These contactees/abductees came from a wide variety of occupations, cultural backgrounds, and age groups.

We employed private detectives and medical doctors, as well as exhaustive hypnosis sessions, in an attempt to learn what archetype had been fed into their particular group consciousness. We never found any implants that were detectable to X-rays, but our hypnotic sessions turned up an incredible amount of fascinating, albeit bizarre, information about underground UFO bases, hybrid aliens walking among us, and thousands of humans slowly turning into automatons because of readjusted brain wave patterns.

During a six-month period, we encountered as many as twenty-five abductees who gave us the same word-for-word accounts of having received an implant. In some cases we were told beeping sounds would indicate that a message was about to be received or a UFO observed.

Since we had determined that the implant was not physical by our earthly definition, what was it? If it was purely a psychological aberration, how could it be that so many men and women were sharing the same delusion, fixation, or fantasy?

Some of these individuals were totally unimaginative people. They read nothing to speak of beyond newspapers and occasional news magazines. To suggest that they were avid science fiction buffs would be preposterous. Most of them were not buffs of any field of human endeavor. How was it such unspectacular individuals were selected as abductees/contactees in the first place?

The accounts came in at a dizzying rate of speed. One that I received was from an intelligent woman in Oregon who had originally maintained an extremely skeptical attitude toward such matters but who had had the following experience:

"Just before sleeping, I became aware of strange sensations at the base of my skull. There were two beings around me, and I realized that they were operating on my head. I felt the movement of instruments, but only slight discomfort, and I knew that something was being implanted. Immediately upon becoming aware of all this, I was put to sleep. A few days later, during a reading by a psychic, I was startled to hear him say, 'You've just had an operation on your head. Something was put into your skull. Something like a radio receiver. It is for receiving space communications.'"

The Oregon report is very typical of those who claim to have received implants. There is an initial awareness of aliens and strange sensations at the base of the skull. Once the implantation has been accomplished, the contactee usually undergoes a kind of cosmic-consciousness peak experience that involves a personal transformation. Many of the contactees subsequently complain of occasional periods of discomfort, followed by heightened awareness. Most mention occasional beeping sounds in the ear, but they also attest that the irritating noise serves as an alert to extended contact.

One contactee reported that when he hears a beep in his ear, he is able to step out of his house and observe two UFOs overhead. Two well-known contactees, one from British Columbia, the other from Massachusetts, indicated that similar sounds in their ears informed them when they could trip their camera shutters and take photographs of UFOs or UFOnauts.

Since we could easily demonstrate that no actual electronic instrument had been implanted in the skulls of the contactees, just what was taking place? If we are not talking about physical additions or adjustments, might we be confronted by the manipulation of certain psychic or energy centers of which we know very little at our current level of physiological knowledge?

In comparing notes with a fellow researcher at a UFO conference, we found that we had each discovered a number of contactees for whom the UFO experience had begun back in very early childhood.

There is something almost primeval in the alleged examinations aboard UFOs. In those primitive societies that emphasize pubertal initiation or rites of passage, the young child or the supplicant is snatched away by masked members of a secret society or by stony-faced elders. They are taken to a place of testing, often unconsciously womblike in design or shape. When they have endured the testing process, they are returned to the village a new person, a transformed individual.

So, too, is the UFO abductee taken away by secret people whose faces, with their large, expressionless eyes and slit-like mouths, appear masklike. The place of testing is not only womblike, but an egg-shaped enclosure. And after they survive the rites of initiation, the abductee is returned to society as a transformed individual.

It is possible that beings from other worlds or dimensions are surreptitiously entering our plane of being and examining our species for purposes of their own. On the other hand, it may be that the abductees are having genuine encounters, but of a nature whose true and total significance we have yet to determine. Is humankind continuing to relive a cultural memory of trial and initiation?

Or are we the unwilling subjects of interdimensional alien experiments?

FOUR ABDUCTIONS FOR BARBARA

Dr. R. Leo Sprinkle of the University of Wyoming in Laramie hypnotized and regressed Barbara Schutte of Wever, Iowa, and discovered that she had been a four-time victim of UFO abductors.

It was in September of 1981 when Barbara awoke dizzy and nauseated with strange bruises on her body and decided that she would seek help to explain what bizarre occurrence had taken place during the night.

She had been troubled by a vague memory of someone holding her down and examining her mouth. When she consulted Dr. Sprinkle, it was determined under hypnosis that she had been escorted to a small room with bright lights. She had been placed on a soft couch by abductors who appeared to her like little children. "They make me smile," she said. "I don't see any ears. They have no hair. Their eyes are black. They have baby-soft skin, lighter than ours."

She next found herself in another room. She was now frightened, and she felt a sharp pain in her back. A tall alien that reminded her of a soldier, with a V-shaped device on his uniform, was with her.

Her abduction, Dr. Sprinkle determined, had taken place in April 1973, when Barbara was twenty-two years old. In a second hypnotic session, Dr. Sprinkle learned that Barbara had first been abducted when she was only eight years old.

The encounter that occurred in 1981 had been the one that had brought her to Dr. Sprinkle, and it had taken place while she had been lying in bed. She had heard something outside, and she remembered feeling frightened. The next thing she knew, she was being taken outside to a derby-shaped craft with flashing lights. She remembered undergoing an examination and being dealt sharp pains in her leg and her head. When she awakened, she was in her own bedroom, but there were strange, unexplained bruises and marks on her body.

In February of 1982, Barbara underwent her fourth UFO abduction. During a hypnosis session by Florida UFO investigator John Blair, it was determined that Barbara had been "floated" from her home to a spacecraft, given tests by aliens, and injected with some kind of fluid.

Dr. Sprinkle has speculated that there may be hundreds of thousands of people who have had a UFO encounter but who were not even aware of it at the time. Dr. Sprinkle lists several characteristics common among people who have had such experiences:

1. An episode of missing time. Under hypnosis many people remember driving down the road and then being back in their car. They know that *something* happened between the two points of consciousness, but they can't fill in the missing time.

2. Disturbing dreams. The abductee will dream about flying saucers, about being pursued and captured, and about being examined by doctors in white coats.

3. Daytime flashbacks of UFO experiences. While they are doing tasks in their normal daytime activities, abductees will flash back to some kind of UFO image or UFO entity.

4. Strange compulsions. Dr. Sprinkle told of one man who, for seven years, felt compelled to dig a well at a particular spot. Under hypnosis he revealed that a UFO being had told him they would contact him if he dug a well.

5. A sudden interest in UFOs. The abductee may suddenly give evidence of a compulsion to read about UFOs, ancient history, or pyramids and crystals, without knowing why.

Dr. Sprinkle advises that if anyone has any of these symptoms, hypnosis is one of the very best ways to uncover one's memory of such an encounter.

ONE OUT OF EIGHT UFO WITNESSES MAY BE AN ABDUCTEE

David Webb, an Arlington, Massachusetts, solar physicist, cochairman of the Mutual UFO Network (MUFON), a top UFO research organization, believes that space aliens have abducted one out of every eight people who has reported seeing UFOs. "Most of these on-board experiences are abductions," Webb stated. "The individual does not want to go aboard the spacecraft, but they are somehow forced into it by the aliens."

In Webb's research, in many cases the victims undergo some kind of examination, but they usually remember nothing of the on-board experience. "Usually they suffer some sort of amnesia, but a lot of information can be retrieved under hypnosis, which often confirms a missing period of time. The person is driving along a road, suddenly comes to a point further down the road, and finds that one or two hours have passed without explanation."

A PANEL OF ABDUCTEES SHARE EXPERIENCES

The abductees speaking at the MUFON's Washington, DC, conference in June of 1987 reported frightening and disorienting aspects of their UFO experiences. They said that they often remembered the events only in fragments and flashes until they underwent hypnotic regression. For the abductees speaking on the panel, the interaction with the UFO entities had seemed primarily to be negative. They told of the frustration of being partially paralyzed and taken without their consent to undergo medical examinations.

A woman in her mid-thirties said, "In a way, you can't blame the aliens for taking us. It would have cost them millions of dollars to get volunteers. If they had asked me, I definitely would have said no. The thing that makes you angriest is that they don't care whether you want to go with them or not. They don't seem to have any understanding of the fact that we have a sense of free will here on this planet and that we think and act as individuals."

As reported by Vicki Cooper in *UFO Magazine: An International Forum on Extraterrestrial Theories and Phenomena*, volume 2, number 3 (1987), the woman went on to say that much of the anxiety of the abductee experience comes from the culture shock that it provokes. "You have to reevaluate your value system because suddenly there is no longer just a possibility that there are other life-forms, but a great probability that they exist—because you've met them. And not only have you met them, but you're not able to be equal with them because they're calling the shots."

Another panelist, a lawyer in his forties, said that there was also the disturbing fact that no one would believe the abductee. "The major part of our problem is that because of the particularly extraordinary science fiction nature of the cause of the fear, we don't have the usual support system of friends and relatives to fall back on. And there aren't an awful lot of psychotherapists or psychologists who have specific training to be able to deal with this kind of trauma."

The attorney went on to say that he was fortunate in that his experiences were rather benign. He received a feeling of reassurance from the space beings. He stated that he had undergone about twenty encounters with extraterrestrials—only one of which involved the classic ETs with the large heads and the large, almond-shaped eyes. The rest of his encounters had been with entities that looked just like normal human beings except that they did not speak. They used telepathy. They gave him the feeling, "Hey, we're glad to see you. Everything is going to be fine, but don't tell anybody about us."

Kathie Davis, the woman profiled in Budd Hopkins's *The Intruders* as having been the unconsenting mother of nine alien babies, reported that she had dealt with her experiences by denying them. "I can live with it because I don't believe it. I really don't. I mean, I'm from the Midwest. Maybe I could accept it more if I was from New York."

Reporting her experiences nervously, but with a sense of humor, Ms. Davis went on to say, "I don't know why I'm afraid, really. I haven't been injured. I don't think it's a terrible thing, and I wasn't left with a lot of anger. I didn't think it was a spectacular thing or exciting, either. I had more anxiety than anything else."

Whitley Strieber, author of the bestselling book *Communion*, said that he had attempted to deal with his tension and anxiety over having undergone an abduction experience by writing about his encounter.

Strieber told the group assembled for the abductee panel at the Washington, DC, conference that when he first realized what had happened to him, he was suicidal. He said that he would have preferred just jumping out a window. Then he began to investigate some UFO literature and discovered that others had had similar experiences. He sought out the services of a hypnotist, thinking that perhaps that would be the last of the ordeal. It wasn't, of course, and he wrote the book *Communion*, thinking that surely now the memories and the feelings would go away.

"I don't think anything will make them go away," Strieber said. "Whatever is happening or whoever is here is here to stay and is here to stay for every one of us."

Strieber went on to say that he had received thousands of letters from other abductees. "They are people, almost to the letter, who are afraid of publicity, including entertainers, political leaders, and members of the armed forces in high positions."

Strieber said that all of these abductees had reported a basic progression of emotions, moving from uneasy, fragmented recollections to a clear memory accompanied by fear. If the abductees consented to undergo hypnotic regression, they usually became even more terrified. Instead of attempting to glean more and more information about the abductee through hypnotic regression, Strieber suggested that concerned researchers should be trying to help these individuals with their fright.

"We need to start learning to get beyond the fear to the stage of insight that comes from the emotional state of coping," Strieber declared. "Beyond that, the ultimate stage, which probably no one on Earth has reached yet, would be true understanding. Once we reach that stage, we will be completely empowered, and none of the abductees will be picked on anymore."

Strieber admitted that reaching such a stage of understanding is a slow process because of the involuntary panic that accompanies the abductee ordeal. "The last time I had an experience with the entities, I tried hard to take a camera and tape recorder, but I couldn't," the author revealed.

When members of the audience asked the panelists why they felt they had been chosen for abduction, none of the abductees voiced that they had any idea. All were concerned with the theory that abductions seemed to occur over several generations of the same family, and they all indicated that they did not wish to have their children or their future children become a part of the abductee experience.

THE MYSTERIOUS PROCESS OF SELECTION

Over the past two decades, hundreds of people who claim the abductee experience have contacted me, and I have often speculated about the process of selection involved in these alleged acts of cosmic kidnapping.

Why are some men and women selected for what would appear to be authentic visionary experiences while the great majority of percipients must settle for glimpses of lights in the sky?

Why do certain witnesses interpret the interaction with UFO intelligence to be an enlightening one while others regard the occasion as one of horror and pain?

And there are other instances in which it is difficult to determine whether the percipient has somehow failed some kind of "test"—which would have qualified him or her as an intergalactic guinea pig—or whether the percipient has been singled out for some kind of extended observation.

JO ANN FAILS HER UFO PHYSICAL

It was 1966. In her boardinghouse in a small Arkansas community, Jo Ann had taken in a roomer named Ernie who was a quiet fellow but vague as to his personal background. Although she is an unassuming person, definitely not the nosy type, Jo Ann still felt a strangeness about the man. But she asked no personal questions, and he volunteered no information. He got a job and made a down payment on a truck.

One night after Ernie had moved from their boardinghouse, Jo Ann had what she thought was a dream in which she saw Ernie come to the foot of her bed and motion her to follow him. Without awakening her husband or her son, Jo Ann arose, put on her housecoat, and went with him. Before leaving the house, he apologized but insisted that he must blindfold her.

Ernie drove for miles before he stopped. He then led Jo Ann through weeds and up some steps before he removed her blindfold.

Jo Ann said that she was aboard a spaceship and was in a kind of laboratory. Entities around her wore long dark cloaks and hoods, but their faces were indistinct.

Jo Ann was placed on what seemed to be an operating table. Wires were attached to various parts of her body.

After she had been examined, one of the creatures said to Ernie, "Didn't you know that this woman has had surgery so that she cannot bear any more children?"

Ernie answered: "No. I came to know her and her son. I thought that she had the potential to be a good specimen for artificial insemination. I thought she would be an excellent mother for our child."

Jo Ann was immediately released from the table. She was once again blindfolded, walked back to the car, and taken home. The next morning she awoke thinking about her strange but vivid dream.

Jo Ann's son, meanwhile, was a member of the Civil Air Patrol Cadet Program. A few days after her strange dream, he was called to an airfield for a meeting. When he returned home, he told her that there had been a group of government men checking a circular burnt area on the other side of the airfield. The officials had put the youths to checking the surrounding area for anything they might find.

The investigators discovered where a car had been driven onto the field, parked, and left. The night of Jo Ann's "dream," a rental car had been taken from the agency at the airfield (no one knew how the key had been removed from the hook *inside* the locked building). The car had been replaced with twenty-five miles of usage registered on the odometer.

Jo Ann also learned that Ernie had disappeared from town that night. His truck was left abandoned on a city street until the police towed it away. When she checked three years later, she found out that the truck had never been claimed and had at last been disposed of by the police.

A BEAUTIFUL SPACE BABY

Mrs. Cynthia Appleton of Birmingham, England, appears to have passed her physical and become involved in a cosmic plan to develop a hybrid race of UFOnauts and earthmen. Mrs. Appleton's spaceman manifested right in her living room, "like a TV picture on the screen: a blurred image and then suddenly everything was clear."

The spaceman told Mrs. Appleton that a year later she would have a "space baby." Although she protested that she was not even pregnant, the "Venusian" told her that the child would be a boy with fair hair, that he would weigh seven pounds and three ounces, that he would be born in late May, and that he would be called Matthew.

The spaceman proved to be correct in every detail with the exception of the date. Matthew was born two minutes after midnight on June 2. Mrs. Appleton was informed, however, that her worry over the date may have delayed the delivery.

Mrs. Appleton told newsmen: "Matthew has a lovely Sun look. He is almond colored allover, and not a blemish. Of course, my husband Ron is the father, but really, the baby will spiritually belong to a race who live on Venus."

THE PSYCHOKINETIC "SPACE KIDS"

In 1976, Dr. Andrija Puharich and his associate Melanie Toyofuku told me of their research with the "space kids" who were sprouting up all over the world with demonstrable psychokinetic abilities and extremely high IQs. Puharich said that these boys and girls very straightforwardly claimed to visit spaceships in their astral bodies.

"The funny thing is," Puharich said, "when two of them meet in a spaceship, they start swapping notes. It's really funny, and they're very cool about it."

Puharich told of an interesting experience that he had had in Mexico. He gathered six of the space kids and started teasing them with some equations and symbols that he told them "aren't known on Earth."

He wrote some things out and asked the kids if any of them recognized the equations. "Yeah," said one of the kids, "but you didn't draw it right. There's a little thing that should go here!"

"Immediately the kids got into it," Puharich told me. "In one half hour—and I have all this on tape—they'd gone through the various progressions.

"When I asked them later if they'd ever thought about these problems before, it came out that they had not. But somehow they remembered it, either from these classes aboard spaceships or preprogramming and getting it all together!"

Are all of these experiences real? Did they actually occur? And if so, why are they occurring?

Whatever else these experiences indicate, they present serious investigators with a cosmic jigsaw puzzle consisting of strangely formed pieces.

CHAPTER TWO

Close Encounters of the Frightening Kind

A young man on his way home from his girlfriend's house experienced what must certainly be one of the favorite little tricks that UFO intelligences employ in their bizarre games with humankind. According to twenty-one-year-old Leon P. Gaines of Peoria, Illinois, a UFO took control of his automobile on September 24, 1966, and drove him several miles out of his way.

Gaines first noticed the UFO when he was taking Miss Geneve Robinson to her apartment. Both Gaines and Miss Robinson described the object as large and blue, and they both testified that the object had hovered above Gaines's automobile for a while. When the object lowered itself to a few feet above the ground, Gaines drove his car toward it to investigate. The UFO allowed Gaines to come within a few yards, then it zoomed silently into the sky and disappeared.

Startled to witness a UFO at such close range, Gaines and Miss Robinson spent several minutes discussing just exactly what they had seen and comparing their respective impressions of the object. Impressed as he was with the initial sighting, Gaines truly had seen "nothin' yet."

A short time later, as Gaines was driving home from Miss Robinson's apartment house, the round blue object suddenly reappeared.

"It just seemed to take control," Gaines said. "My car began to pick up speed . . . the brakes wouldn't work . . . the doors wouldn't open."

Gaines tried desperately to steer his automobile, but his pressure on the wheel had nothing to do with where the vehicle actually went. The UFO stayed above Gaines's car for several miles, directing the automobile's course and speed. When the young man was at last able to regain control of his automobile, he drove directly to the police station to report the incident. Gaines felt that the officers might reject the idea of charging a traffic violation against a UFO, but it made him feel better to tell someone about the harrowing event.

"This kind of thing really shakes a guy up," Gaines said in something of a classic understatement.

HUMANOIDS IN A WASHINGTON BEAN FIELD

On August 13, 1965, two Renton, Washington, sisters came to work at 7:00 A.M. to get an early start in Yas Narita's bean field near Kent. Ellen and Laura Ryerson had barely entered the bean field when they noticed that three "workers" were already walking in the area. The teenage sisters had not been in the field long, however, when they discovered that the three strangers were more interested in them than in gathering beans. Even more frightening was the girls' discovery that their three fellow bean pickers were not human beings.

The three strangers had white-domed heads and protruding eyes. They were between five feet, two inches and five feet, five inches tall. The flesh of their expressionless faces had very large pores, and their complexions were "gray, like stone." The three aliens wore sleeveless purple V-necked jerseys with white shirts underneath.

Fortunately for the girls, the three strange "men" were easily outdistanced and appeared to be without weapons. Ellen and Laura were able to get back to their automobile, and they sped away to make their report to the Washington State Patrol.

ALIENS AT THE SIDE OF THE ROAD

It was about 10:45 P.M. EST on March 20, 1967, when a man—pseudonymously dubbed "Mr. Rible" by Robert A. Schmidt, then secretary of the Pittsburgh UFO Research Institute—asked his daughter Jean to accompany him in the family Volkswagen to the outskirts of Butler, Pennsylvania, in the hope of glimpsing some unusual aerial light phenomena that he had been observing. Since they lived only a mile from a private airfield, Rible felt rather strongly about his ability to distinguish conventional from unconventional lights in the sky.

They parked the Volkswagen on a back road and, after a few minutes' wait, spotted two globes of light. The illuminated objects appeared at first to be two airplanes flying parallel to the highway, toying with the notion of landing on the road. A short time later they gave the appearance of having done just that.

Then, from a distance of about a quarter of a mile away, the vehicles came up the slope toward the Ribles at a speed of about eighty miles per hour. The Ribles, who had stepped out of their automobile for a better look at the globes of light, now prepared themselves for what seemed to be an inevitable collision with the two berserk aircraft.

The crash never came, but the Ribles were confronted with an impact to their construct of reality that challenged them—and all of us.

The lights seemed to transmogrify into a semicircle of five figures, who stood just a few yards from the hood of the Ribles' Volkswagen. Both Ribles jumped back into the car, but while her father worried over starting the vehicle, Jean got a good look at the figures. Schmidt quotes her as describing them in the following manner:

"They looked like human beings, but their faces were totally devoid of expression. Their eyes, if you could call them such, were horizontal slits. I could not see any irises or pupils—just slits. Their noses were narrow and pointed, not unlike a human nose, and their mouths were slits like the eyes."

Jean said that four of the figures were about five feet, seven inches in height, while the fifth was about five feet tall. They all wore a kind of flat-topped cap, with blond hair showing to ear length. The shorter humanoid had hair to the shoulders, which caused Jean to suspect that it might be a woman.

All five of the beings were dressed alike, in sloppy gray-green shirts and trousers. The skin on their faces and hands was rough-looking, resembling "scar tissue or skin which has been severely burned."

Jean admitted that the semicircle of staring entities gave them "the creeps."

"We heard no noise in connection with either the lights or the figures," she told investigators. When the engine was started, the Ribles had to "reverse and then go forward and round the figures to miss them."

Under persistent questioning designed to ferret out details of the experience that might have been forgotten under stress, Jean remembered what may have been a most significant factor. As the lights swiftly approached their car, she heard a "chorus of voices" in her *head*, not her ears. She seemed to sense them with her brain.

The voices said: "Don't move . . . don't move . . . don't move . . ."

They kept repeating, "Don't move . . . don't move," but they dragged it out—"Dooooooonnnn'tttt moooooove."

"When the lights vanished, the voices stopped at once," Jean said.

SNATCHING AIRPLANES FROM THE SKIES

UFOs chase airplanes, too. What is worse, sometimes they catch them.

In the May 1961 issue of *Fate* magazine, Tom Comella, an active UFO researcher, related his interview with Master Sergeant O. D. Hill from Project Blue Book. Comella's article was entitled "Have UFOs Swallowed Our Aircraft?" He quoted Sergeant Hill as saying that it was the mission of the Air Technical Intelligence Center at Wright Field to "prevent another Pearl Harbor." When the sergeant was asked if it were true that aircraft had been disappearing, he replied: "I must confess to you that is true; our planes have and still are disappearing from the sky."

According to Comella, who claimed to have quoted Sergeant Hill verbatim in the article, the representative from Project Blue Book related two cases of mysterious aircraft disappearances, deleting only such classified information as dates and places.

The first case dealt with an air force F-86 jet fighter sent aloft to investigate a UFO that had been plotted on the radar scope of an air base. The radar station had the F-86 on its screen as it circled the field in search of the UFO. Then, according to Sergeant Hill, "the radar operator spied the UFO heading straight for the craft. The operator radioed the jet to climb at once, but it was not fast enough. Before the eyes of the radar operator the two blips merged on the screen. *Then there was only one.*"

The radar operator attempted to establish communication with the blip that was then speeding away, but there was no response from the UFO. The F-86 was never seen again. The combined efforts of air force and army search crews were unable to uncover even a shred of wreckage. Sergeant Hill said that the air force had classified the case as a "mysterious disappearance."

The researchers immediately brought up the Kinross case, which had been made public by Major Donald Keyhoe. In this incident, an F-89 jet interceptor was dispatched from Kinross Air Force Base in Michigan after Air Defense radar plotted a UFO over Lake Superior. As in the case recounted by Sergeant Hill, two blips had merged into one and the UFO had shot away toward Canada. Nothing was ever found of the jet or its two-man crew.

Sergeant Hill's second case concerned the UFO kidnapping of a transport plane with twenty-six persons aboard. The transport was being carefully tracked by a radar station and was maintaining constant radio communication with the base when the operator suddenly radioed the transport's pilot and advised him to be cautious of an unidentified object that was rapidly moving up on the aircraft.

In the words of Sergeant Hill: "The UFO was traveling at a high rate of speed, about two thousand five hundred miles per hour. It jumped about on the radar scope like a tennis ball. All of a sudden, the mysterious blip headed straight for the transport plane, and before the radar operator could warn it, the two objects had united into one on the radar screen. The one remaining blip sped straight up at a terrific rate of speed. A surface search of the water in the vicinity revealed no oil slick, although a general's briefcase was found floating around. The plane had completely disappeared!"

At 7:06 P.M. of October 21, 1978, Frederick Valentich was on a short solo flight from Melbourne, Australia, to King Island in his single-engine Cessna 128 airplane.

Spotting four bright lights above him, the twenty-year-old pilot radioed air traffic controllers on the mainland, asking if the lights might not be the landing lights of a military aircraft. He was told there were no military planes in the area.

At 7:08, Valentich reported that the lights were approaching from due east toward him and that they seemed to be playing some sort of game, flying at a speed that he could not estimate.

At 7:09 he said, "It is not an aircraft, it's—"

Radio contact was broken for a short time. When it was reestablished, the young pilot was saying, "It is flying past. It is a long shape. I cannot identify more than that. It's coming for me right now. It seems to be stationary. I am orbiting and the thing is orbiting on top of me. It has a green light and a sort of metallic light on the outside."

There was a brief silence, and the pilot radioed that the Cessna's engine was experiencing coughing and rough idling. His final appraisal of the mystery object was, "It is not an aircraft."

That was the last that anyone ever heard of Frederick Valentich, for after uttering that pronouncement, he vanished.

For four days, through Wednesday, October 25, a land, sea, and air search sought some kind of evidence of the fate of the young pilot. The only item of substance that the searchers were able to locate was an oil slick that investigators subsequently concluded could not have been left by a small airplane.

The official explanation was that Valentich had been flying upside down and had perceived reflections of the King Island and Cape Otway lighthouses in the clouds. However, as pilots and other aviation authorities pointed out, a Cessna 128 cannot be flown upside down.

According to certain sources such as the Australian national newspaper, there had been no break in radio transmission at all, and Valentich had described a massive UFO in detail to the traffic controller. Guido Valentich, Frederick's father, stated that he had been told that the tapes contained a lot more about the UFO and that his son's voice was calm right through the final transmission.

Guido Valentich was quoted as saying that he believed Frederick may have been "borrowed" by visitors from another planet, and he prayed for the return of his son. In the absence of any other evidence, the bereaved father stated that he believed Frederick may have been abducted by aliens.

The young pilot's girlfriend told reporters that she intuitively knew that Fred was still alive and that she would see him again one day soon. Rhonda Rushton claimed to have had a "top secret" interview with officials. According to Miss Rushton, "We all know Fred is alive, and we have told the authorities this. But it is still all top secret now. I will not be saying any more about this."

A spokesman for the Department of Transport denied that any part of the taped radio exchange had been suppressed. The spokesman went on to state that the text of the radio messages between the aircraft and the ground began when the pilot requested the flight service unit at Melbourne Airport to provide information on other traffic below five thousand feet. The transmission finished about six minutes later at 7:12 P.M. when Valentich's aircraft did not respond to further calls from the ground.

In an attempt to quell rumors that they were squelching information, the Department of Transport released what they claimed was the complete transcript of Valentich's final communications with traffic control.

7:06 P.M.   *Pilot to ground:* Is there any known traffic in my area below five thousand feet?

Flight Service Unit: Negative. No known traffic.

Pilot: Seems to be a large aircraft below five thousand feet.

Ground: What type of aircraft?

Pilot: I cannot confirm. It has four bright lights which appear to be landing lights. Aircraft has just passed over me about one thousand feet above.

Ground: Is large aircraft confirmed?

Pilot: Affirmative. At the speed it is traveling, are there any RAAF [Royal Australian Air Force] aircraft in vicinity?

Ground: Negative. Confirm you cannot deny aircraft.

Pilot: Affirmative.

7:08 P.M.   *Pilot:* It's not an aircraft. It's . . . [break in transmission]

Ground: Can you describe aircraft?

Pilot: It's flying past. It has a long shape. Cannot identify more than that. Coming for me right now. It seems to be stationary. I am orbiting and the thing is orbiting on top of me. It has a green light and sort of metallic light on the outside. [The young pilot then told Ground Control that the UFO had vanished.]

Ground: Confirm that it has vanished.

Pilot: Affirmative. Do you know what sort of aircraft I've got? Is it military?

Ground: No military traffic in the area.

7:12 P.M.   *Pilot:* Engine is rough, idling and coughing. Unknown aircraft now hovering on top of me.

Ground: Acknowledge.

There was a long metallic noise, and contact with Valentich's aircraft was lost forever.

Between 1959 and 1961, dozens of nervous candidates for Soviet civilian flying licenses complained about UFOs swooping at them and even following their planes back to their airfields.

A number of large, cigar-shaped "mother ships" were sighted in Russian skies, and their lengths were estimated at from eight hundred meters to one kilometer. At night these huge vehicles gave off an intense bluish-white color. In daytime and when they were stationary, the mother ships were most often described as being grayish in color.

In 1961, near Irkutsk, a doctor photographed a UFO on the ground. She was also able to photograph two saucer occupants that emerged from the spacecraft.

While on a regular scheduled flight across the central plains of the Soviet Union in 1961, a mail plane with four passengers on board was snatched from the skies by a UFO. According to Italian science writer Alberto Fenoglio's informants (Robert Pinotti translation): "The machine was found, intact, two days later, near Tobelak [Siberia]. Everything on board . . . was in perfect order. The tanks contained fuel for two hours of flight. *The four passengers had vanished without a trace.*

"At a distance of one hundred meters from the aircraft there was a huge, clearly defined circle, thirty meters wide, on which the grass was all scorched and the earth depressed. A 'flying saucer' had landed there."

Even more bizarre than an airplane kidnapped from the skies was the story of a waylaid parachutist. In *Olthe II Cielo-Missili E Bazzi*, Fenoglio writes: "In 1961, a woman parachutist jumped from a height of nine thousand meters. The pilot of her aircraft saw her floating down, with her parachute extended, so he landed to wait for her. She came down . . . at Saratov . . . *three days later.* Her explanation? She had been caught in midair by a 'saucer.' Its three occupants had treated her well, had taken her out to an immense distance in space so as to view Earth, and had given her a message for the authorities. The envelope containing the message was given to the local chief of police. We are told nothing as to its contents."

A GAME FOR WIZARDS

Bill Fogarty of Phoenix was a twenty-year-old junior at the Indiana University South Bend when he saw a UFO in April 1972. He had been a member of an informal group that got together once a week to discuss politics, philosophy, and other subjects.

One night someone had brought up the topic of UFOs—and that very evening, driving back to their respective apartments, Fogarty and four friends claimed to have witnessed a low UFO overflight.

Fogarty assured me that the five were all college students, physically fit, nondrinkers, nondopers, and two were Vietnam combat veterans. Each of them prided himself on maintaining a cool, analytical approach to all aspects of life, especially toward anything that smacked of the occult or the bizarre. And yet each of them swore that he had seen what was unmistakably an object in the sky that he could not identify as a conventional aircraft, an ordinary celestial manifestation, a weather balloon, a bird, or anything known that could have been flying above them.

Four nights later, two of the five had seen another UFO. Then, on the next evening, Fogarty and the other two saw a brightly glowing object overhead as they returned at around one o'clock from a movie.

It was not long before the five of them had a group sighting, and from that evening on they had taken to nightly sky watches.

"We all witnessed UFOs cavorting in the midnight sky," Fogarty said. "On one occasion I stood within ten feet of two nocturnal lights hovering silently in midair. Later we heard rappings in the dark, hollow voices, heavy breathing, and the crushing footsteps of unseen entities.

"Strangely enough," Fogarty said, smiling, "we were able to maintain our cool toward all the phenomena occurring around us. Maybe we got to thinking that we had been chosen for some special kind of interaction. Perhaps, secretly, we were beginning to view ourselves as masters of two worlds. I mean, we were all Dean's List students, all athletic young men, normally balanced emotionally, mentally, sexually. I guess we felt that modern Renaissance men such as ourselves could deal rationally with such phenomena and stay in control of the situation."

But then, Fogarty went on, the manifestations became violent. They swept through the home of one group member one night, pounding on walls, yanking furiously at the bedposts, striking the startled young man in the face, terrorizing his entire family. Some of the group were followed by unmarked automobiles that seemed a bizarre mixture of styles and models—phantom automobiles, if you will.

Within the next few months, the number of harrowing incidents increased and expanded to include strange, dark-clad, nocturnal visitors in the apartments of several group members.

Radios and television sets switched on by themselves. Doors opened and closed—although, when tested, they were found to have remained locked. One of the group made the wild claim that he had been teleported one night from his bedroom to the middle of a forest on the outskirts of the city.

"As preposterous as that sounds," Fogarty said, "I'm sure that most of us accepted it as true, since we had all undergone some incredible experiences. We had all lost our sense of perspective. I began sleeping with the light on and a .38 Special under my pillow. Another of my friends invested heavily in weapons and began running with a group that offered sacrifices to Odin. A third was 'born again' into fundamental Christianity. The other two dropped out of college a month before they would have graduated with honors.

"I've thought a lot about what all this meant," Fogarty said. "I think the five of us had entered a kind of game, a contest, a challenge, a testing experience. The trouble was, we just didn't know all the rules.

"Modern society doesn't prepare us to play those kinds of games. Modern society doesn't tell its kids that there is another reality around them. Our educators have ignored the individual mystical experience and the other dimensions that can open up to those who enter altered states of consciousness—whether it be through drugs or through accidentally stumbling into the twilight zones.

"I had the feeling," Fogarty recalled, "that my friends and I were dealing with some kind of energy. At first I thought it was something from outer space, some alien world. But I've thought out our experiences a great deal over the past eleven years, and I believe that we somehow activated some energy that is a part of this planet. I think we might have triggered some kind of archetypal pattern with our minds. Maybe that's what magicians have tried to do since Cro-Magnon days—interact with and control that energy with their minds."

How had such intelligent, resourceful young men lost control? Why had they ended up paranoid, frightened, or converted?

"Because we weren't magicians, obviously." Fogarty chuckled. "We had no idea just how deadly serious the game could become. It really is a game for wizards, not for some smart-ass college students who believe their brilliant intellects and their physics books can provide an answer for everything."

FOUR FEARFUL ENCOUNTERS FOR THE SPANISH TRUCK DRIVER

"Maxi Iglesias does not seem very imaginative. If he is lying, he does it to perfection."

Thus stated reporter Angel Gomez Escorial, writing in *White and Black*, a Spanish magazine, about a young truck driver who reportedly witnessed UFOs on four separate occasions and had two close encounters of the frightening kind.

On the night of March 20, 1974, twenty-one-year-old Maximiliano Iglesias Sanchez was driving his truck past the village of Horcajo on his way back to Lagunilla when he noticed a very strong white light some two thousand feet ahead on the highway. At first he assumed it was another truck, or perhaps a car. He switched his headlights to high beam several times to signal the other driver to dim his own lights. The bright light remained almost blinding in its intensity. Its brilliance forced Maximiliano to pull his truck to the side of the road.

The bright light eventually dimmed to about the power of a fifty-watt bulb. Maximiliano then continued to drive toward the light. When he was about six hundred feet away, Maximiliano discovered that the thing was indeed something *very* strange: Without warning, all the lights on his truck went out and the motor stopped. The area was illuminated only by the now-dim light of the craft.

Sanchez described the object as having a metallic structure of either platinum or steel. It seemed quite solid, and it had smooth edges without rivets or openings of any kind. It was thirty to thirty-six feet in diameter and rested about five feet off the ground on three round landing pads.

"It was a light like I have never seen before" was how Sanchez described the dim light, which seemed to be uniform on all surfaces of the UFO.

He then noticed a second, similar "ship," as he called it, above and about fifty feet to the right of the first.

As if from nowhere, two beings appeared in front of the grounded UFO. They moved together and began motioning to each other "like the tourists do." They looked at Sanchez, and one pointed at him. At that, one of the humanoids turned around and disappeared to the right of the first UFO, while the other remained to watch the young trucker.

Soon the other being returned. The two entities looked at each other; then both disappeared to the right of the ship. Soon it slowly lifted into the air with a slight humming sound.

Sanchez described the humanoids as about six feet tall and wearing close-fitting coveralls. The material of their coveralls was brilliant, like the ship, and appeared to be made of a rubbery material. The occupants' walk was "normal," not like that of a robot, and their arms and legs seemed to be proportioned like a human's. As hard as Sanchez tried, however, he could not describe their facial features. The encounter had occurred at night, and he was never closer than six hundred feet to the entities.

When the first UFO reached the altitude of the second UFO, the two objects remained motionless in the air. Sanchez then decided to leave. His truck started right away, and the lights worked once again.

But he had driven only a short distance from the twin UFOs when curiosity got the better of him. Sanchez stopped his truck and climbed down from the cab to study them. He noticed that the illuminated ship had dropped down to the site where it had been before.

At this point, and for the first time, Sanchez registered fear. He took off as fast as his truck would carry him, and he drove straight to his home in Lagunilla, where he went to bed immediately without eating.

The following day Sanchez told the story to his neighbors, but he found that they did not believe him. However, the son of his employer told Sanchez he believed that the event had happened, for it was similar to the experience a commercial traveler claimed to have undergone near Seville.

That afternoon, March 21, Sanchez drove to Pineda to deliver a load of construction material. While there, he made his customary visit to his sweetheart, Anuncia Merino.

Sanchez told her and her family what had happened the night before. They insisted that he stay with them for the night. It was getting late, and they were afraid for him to make the trip through that area again.

He did not take their advice, however, but headed back home to Lagunilla.

At about 11:15 P.M., he arrived at the site of the previous night's UFO/humanoid sighting. Once again he saw a bright light ahead of him. Sanchez was convinced that the entities would do him no harm, as on the night before. Again he drove to within about six hundred feet of the light. That night, however, the light was produced by not one but three UFOs.

As before, the truck lights went out and the engine quit—this time with a backfire! One of the UFOs was resting on the highway; the other two were just off the road to the right side of Sanchez, one behind the other. All three were illuminated with the same soft light he had noticed the night before—or, rather, earlier that same morning.

Suddenly four humanoids appeared and walked to the center of the craft resting on the highway. The four looked at Sanchez as though studying him, communicating with one another through gestures. The beings pointed at Sanchez, then started walking in his direction.

Sanchez, fearing the motives of the four, began running along the highway. The four entities increased their pace. Sanchez started cross-country, with the four gradually gaining on him. When he came to a ditch, he jumped into it.

The move to evade his pursuers seemed to work: They had lost him—for the time being, at least. He could watch them from his muddy vantage point as they circled in search of him. Although they were often as near as fifty feet, he could still not make out facial features—a matter that seemed to bother him when he reported the incident.

Finally the four strangers left, and Sanchez felt it was safe to vacate his hiding place.

He started walking, and he was soon within sight of the lights of Horcajo, which he estimated at about a mile or less away. He sat down and smoked a cigarette to calm his nerves. He rested for perhaps ten minutes, then returned to the area where he had left his truck, feeling that by now the spot would be deserted. He was wrong, for the three ships were still there—although he did not see the four humanoids.

As Sanchez reached the truck, something bothered him. The door was closed; he remembered that he had left it open when he departed earlier. His fear that someone might be inside was put to rest, however, when he found nothing and no one in the cab. He tried to start the engine but found it still wouldn't work.

As he shut the door, the four humanoids appeared in the middle of the road, as before, gesturing to one another. They went to the right side of the ship parked there and apparently entered it just before it climbed to an altitude of about fifty feet. The same low humming sound was heard, but it stopped as soon as the UFO came to rest.

It appeared to Sanchez that the UFO was clearing the road so he could leave, just as it had done the night before. The truck started instantly this time, and the lights came on.

"And I buzzed out of there!" he told investigators.

Apparently his survival instinct was not as strong as it had been nearly twenty-four hours earlier, for he stopped the truck about six hundred feet down the road, climbed from the cab, and walked back to the area where the three UFOs were located. The one that had lifted from the highway to let him pass was once again in place on the asphalt paving.

He hid in a clump of bushes about thirty feet from the four humanoids and observed the nearest ship to see if he could find some opening in it through which the beings had been coming and going, but all he could see were unbroken walls.

He watched the beings at work. They were using two tools that resembled a horseshoe and the letter T. They inserted the T into the ground at the embankment of the highway. Then they would withdraw the instrument and insert the horseshoe in the hole. They did not appear to be taking mineral or vegetation samples, however.

Even at close range Sanchez was not able to determine facial features on the foursome.

Not more than about three minutes passed before fear returned to Sanchez, a fear that was stronger than his curiosity. He later reported that the beings never looked in his direction or seemed aware that he was there, but he felt it was time to return to his truck and head for home.

When he reported the incident to his boss the next day, he was advised to contact the Civil Guard, which he did, accompanied by his employer's son.

The officer in charge contacted headquarters in Bejar, and after three days an officer arrived and filed a report, following an interview with Sanchez. Investigators went to the alleged landing site, where they found some strange tracks. On the highway where the craft had landed, the investigators found a deep, straight groove, as if the asphalt had been scorched by a very hard object. On the embankment the investigators found two scratches that seemed to substantiate Sanchez's story about the tools, but this was all the physical evidence the investigators found to indicate that the craft and its strange visitors had been there.

A few days later, two individuals from Madrid arrived in Lagunilla, stating that they were UFO investigators. They were equipped with instruments for making tests, including a Geiger counter. The team, trained in such investigative matters, was successful in finding three circles that appeared to have been caused by the craft resting there. The grass was pressed down, but no indentation from the landing gear could be found. Abnormal radioactivity in the area was recorded by the Geiger counter.

As a footnote, Sanchez added that although his truck started that night, the battery was completely dead the next morning. When he had it recharged, the garage mechanic could detect nothing abnormal about the battery.

The strange-craft sightings did not stop here, however. On March 30, Sanchez was once again with his girlfriend in Pineda. It was 12:45 A.M. when they saw what looked like two large spotlights in the sky at about twenty-eight hundred feet. The spots of very bright light were flying over the area, and they gave every indication of being similar to the UFOs witnessed by him earlier that month.

The fourth and final sighting took place in early May of the same year, while Sanchez was with his girlfriend and her uncle.

Sanchez had gone to the city of Salamanca, his hometown, to take a driving examination for a first-class permit. It was about 6:30 A.M. when Anuncia saw a strong white aerial light, which soon disappeared.

A few miles down the road they saw another bright light—this one coming directly toward them at extremely high speed. Anuncia feared that it was going to strike them head-on, but about three hundred feet before impact, the light changed direction, passed over their car, and disappeared.

Sanchez reported no further sightings of UFOs or humanoids, and soon after these experiences he went into the army. While curiosity at times may have overcome his fear, he was quoted in a radio interview as saying, "There is no need to go on about bravery; before I did not know what fear was, but now indeed I know what it is."

BOLD UFONAUTS IN SOUTH AMERICA

At least two UFO occupant sightings were made in Argentina on June 14, 1968. Pedro Letzel was returning to the motel that he managed when he spotted a brilliant red object near his home. When he entered his house, he found his nineteen-year-old daughter, Maria, lying unconscious on the floor.

Upon being revived, Maria told a most extraordinary tale. She had accompanied some customers to the door and had walked back into the kitchen when she noticed that an intense light illuminated the hallway of the house. Then, as she watched in bewilderment, a being six feet tall appeared before her.

The alien wore a sky-colored suit, which was brilliant in color and scaly in texture. He was blond and extended a celestial sphere in his left hand that moved in all directions. The tips of his feet and hands gave off a strong luminosity. When he moved his right hand, Maria said, he brought about a fainting sensation in her. The draining sensation ceased when he lowered his right hand, which was partially covered by a large ring.

The being seemed friendly enough, Maria testified. He mumbled incessantly in an unintelligible murmur and made noises that sounded very much like laughter, but at no time did he ever move his lips. The alien stayed several minutes, apparently enjoying their strange tête-à-tête, then disappeared. That was when the startled teenager fainted.

At about the same time that Maria was hostess to her uninvited guest, Catalicio Fernandez was entertaining alien beings dressed in brilliant green suits in his home in Buenos Aires.

"I could not understand what they were saying, but their tone was friendly," Fernandez said. "However, I felt dizzy whenever one of these men would raise his arm toward me. When he took it down, I would return to normal. I could see nothing in his hand. There was no ring, signaling ball, or crystal, such as other people who have seen these beings have said."

Fernandez said that his mysterious visitors were about six feet in height and wore clothes closely fitted to their bodies.

"I first noted their appearance when one of them sat down on my bed. When I awakened, he indicated that I should remain calm."

While most people would find it difficult to remain calm if a pair of green-suited strangers suddenly materialized in their bedroom, fifteen-year-old Oscar Iriart accepted an invitation to ride with two red-coated aliens.

Oscar, a farm lad considered diligent and reliable by his teachers and family, was checking a dividing fence on horseback when he saw what he believed to be two hunters waving at him. Except for their sunken eyes and semitransparent legs, Oscar said that they looked very much like two men one might meet any day in Sierra Chica. They were more or less as tall as he (five feet, seven inches) and they engaged him in a dialogue that Oscar later believed had been telepathic.

"You are going to know the world!" they said to him.

"Yes, of course," Oscar agreed. "Someday when I have saved much money, I will travel."

"No," they said. "We will take you now!" Then the beings paused. "No, we cannot take you now after all. We have too much cargo. But we will come back for you."

They indicated an object resting in a deep, muddy ditch as their craft. One of the beings gave Oscar an envelope and told him to stick it into a nearby pond. The boy did as instructed and was startled to note that neither the envelope nor his hand got wet. Satisfied that he was suitably impressed with the demonstration, the beings entered their craft and took off straight up.

"I ran as if in a dream toward my horse," Oscar recalled, "but the animal was completely paralyzed and could not move until the flying object had totally disappeared."

Oscar galloped to his home, told his parents of his encounter with the red-coated aliens, and showed them the envelope. Written on a sheet of paper in crude, heavy lettering were the words: "You will know the world. [signed] Flying Saucer."

The boy's family were extremely concerned with his glassy stare, as if he were just leaving a hypnotic state. They investigated the spot that Oscar had indicated as the site of the UFO's landing, and they were astonished to find three deep indentations that formed a perfect isosceles triangle.

Impressed by this bit of discovery, an investigator named Amarante interviewed the boy a few days later and expressed his opinion that the lad had received "orders encrusted in his brain by telepathic means."

Señora Iriart became nearly hysterical with fear, believing that the UFOnauts meant to return and take her son back with them to another world. Oscar had never read science fiction magazines. He could not have imagined the description of the flying saucer that he gave to investigators. Oscar was a hard-working young man who enjoyed tinkering with old cars and applying himself to his accounting courses.

Police Sergeant Raul Coronel was unimpressed with the story that Señor Iriart told him. He considered absurd and impossible such allegations about alien beings even existing, to say nothing of their returning to take Oscar for a ride.

It was nearly midnight when Sergeant Coronel, Carlos Marinangel, a butcher, Jose Luis Marinangel, Carlos's brother and the administrator of the prison, Hugo Rodriquez, an auto mechanic, and Walter Vaccaro, a member of the Sierra Chica Club, arrived near the spot where Oscar had said that he had seen the object and its occupants nearly twelve hours earlier.

"Look, here," Sergeant Coronel said, punctuating his words by waving a powerful flashlight that he held in his hand. "Even though I do not believe such stories, there are indeed some strange marks and tracks where the boy said he saw the craft."

The other men, with the exception of Carlos, agreed that perhaps the boy should be taken seriously. "You watch too many weird things on television," said Carlos.

They continued to examine the indentations in the soft earth, when suddenly it was Carlos himself who screamed: "Here comes a moving light!"

A short distance away, only a few feet above the ground, a brilliant and luminous object zigzagged its way toward the five investigators. Its movement was slow, as if it were searching for the precise spot where it had been detained earlier that day. As it passed over their heads, the men flung themselves to the ground, fearful of being struck by the object. Sergeant Coronel reached for his pistol, but Carlos Marinangel prevented him from firing at the UFO.

For a few minutes, the brilliant object continued its rather erratic course near the ground. Then, as if concluding that Oscar had failed to keep their rendezvous, the UFO attained greater speed and altitude until it disappeared into the night sky.

Later, before interrogating police officers, the five former doubters kept repeating: "Flying saucers do exist! Flying saucers do exist!"

The most insistent reciter of such a conviction was Sergeant Coronel, who was almost immediately transferred to the regional office of Azul by superiors who had taken a dim view of his declaring the event to the press before having made an official report. Such action, according to his superior officers, was most unlike Sergeant Coronel, who had always been noted for his seriousness of purpose and his responsibility.

From May 29 to June 27, numerous reports of UFOs and their crew members came in from the mountain towns of Peine, Socaire, Tocanao, Chilopozo, and Tilomonte. Residents were said to be terrified of the "invading" disks.

A group of farmers were interviewed by the authorities in the old city of Calama, which grew up around one of the largest copper mines in the range. The farmers testified that the strange flying objects had been seen to land in several areas. Three crewmen had been sighted by local residents. The UFOnauts were described as wearing bright, tailored jumpsuits and were said to approach and stare inside the huts into which they had frightened the local inhabitants.

In Bolivia, during the same period, a long-time member of the police and a young miner claimed to have seen UFOs in the area of El Choro, near Oruro, three hundred kilometers south of La Paz.

The miner, twenty-five-year-old Romulo Velasquez, testified that he had seen a UFO land and had watched a "strange being, tall and thin" emerge and approach him. Velasquez said that the being seemed to want to tell him something, but he was unable to oblige the strange creature. Velasquez fainted.

RARE PHENOMENA IN CHILE

On July 15, *La Razon* carried an astounding report of a series of rare phenomena in Chile:

> *Cauquenes, Chile—Various newspapermen and a group of customs guards are meeting in this region to study a series of strange and inexplicable phenomena that have been occurring for a week.*

> *Wednesday, July 10, on the hill of La Nariz in the coastal zone of the province of Maule, all of the automobiles that traveled there were stopped without reason. Drivers remained in their stalled vehicles all night until suddenly, without explanation, the cars lighted again and began to move. One of the stalled vehicles, to multiply the strange event, was mysteriously drawn up a hill for a distance of forty meters until it stopped next to the road bank.*

> *The latest phenomenon that now seriously preoccupies the thoughts of the inhabitants of Cauquenes occurred last night and was formally announced by the owner of the Curanipe Hotel, Jose Bolosin. According to Bolosin, at 6:30 A.M., before the sun came up, one could feel a heavy heat wave in all the sectors. Fearing that it was a forest fire—because in this zone the usual temperature is below zero—a group of neighbors left to look over the nearby area. A careful examination and search, however, revealed nothing that would explain this unusual*

increase in temperature that reached twenty-six degrees centigrade.

As a witness to the fact, this morning the peach trees of Curanipe appeared completely in bloom, before season, because of the hot wind that has prevailed in the last few hours. It must also be mentioned that residents in Cauquenes claim to have seen flying saucers in the last few days. The facts are being investigated by authorities of the local police.

During this same period of mysteriously stalled automobiles and peach trees blooming prematurely, a woman and her husband traveling near Santiago claimed to have seen an "extraordinary astral apparition." A mysterious, luminous sphere appeared above their car and hovered there to emit a blinding blue fluorescent light, surrounded by a rainbow aura.

"When we came close," the witness said, "the sphere, the size of a bus, disappeared. But when we looked back, it had returned to shine with fluorescent flashes."

UFO OCCUPANTS WITH METALLIC VOICES

Sporadic sightings occurred in Argentina and Chile through the remainder of July and August 1968. Then, on September 2, after numerous citizens witnessed a UFO over Mendoza and two employees of a casino claimed to have confronted UFO occupants, Jose Paulino Nunez, an employee of a distillery, allowed his contactee story to be released.

Enrique Serdoch, chemical technician of the company laboratory, served as spokesman for his friend, who had maintained silence for over a month. Fellow employees Alberto Gonzalez, Roberto Micelatti, Carlos Wengurra, Hugo Torres, Enrique Aporta, and Ricardo Schmid testified to their companion's sobriety and responsibility. They, too, had been aware of Nunez's encounter with the UFOnauts, but they had respected his decision to keep silent about the incident. At the time Serdoch received permission to release Nunez's story to the press, the contactee was in the hospital undergoing surgery, not as a result of his experience.

Here is Nunez's story as related by Serdoch:

"At 1:15 in the morning on the last Sunday in June, Jose found himself on the beach allotted to the analysts. The section is very dark, and the only people who are found there are guards. As Jose came down from a fuel-oil tank, he encountered two people who he thought were guards. He was soon set straight in this regard when the strangers showed him a spherical object some thirty centimeters in diameter, in which colorful figures moved.

"As Jose watched, he saw people who walked about as if they had been filmed by a hidden camera. The depth of the images was clear. The dress and activity of the animated figures within the sphere did not offer anything special to catch his attention.

"Then, enigmatically, one of the strangers asked Jose: 'Do you know these people? They were like you. Many more will be like them. Many people in the world will see the same thing that you have seen. We will talk to you about this again. If you should mention this to anyone, be certain that it is only with responsible people.'"

As Serdoch told it to assembled journalists, Nunez could not explain how he suddenly found himself back at the laboratory.

"What's wrong with you?" Alberto Gonzalez, Nunez's partner, asked. "Why?" Nunez wondered. "What *seems* to be wrong with me?"

"You are white as paper!" Gonzalez told him.

Nunez began to weep, and it was nearly an hour before he could compose himself enough to return with Gonzalez to the oil tanks on the beach where he had encountered the two beings. Nunez said that the strangers' voices had been metallic, as if they had been emitted from the interior of a telephone receiver. Their suits had been one-piece, similar to the kind worn by frogmen.

Authorities in Argentina were kept busy early in September. First the two casino employees, Villegas and Piccinelli, claimed contact with UFOnauts. Then Nunez authorized a friend to tell his contact story.

As if this were not enough, several employees of the Belgrano Railroad reported that the lights of the station house had been mysteriously turned off at 3:30 A.M. on September 1. At the same time, the owner of a Renault automobile was confronted with a similar stoppage of power as he drove in that vicinity.

CAPTURED BY AN ALIEN "HUNTING PARTY"

In the UFO intelligences' weird games of cat and mouse, living people also disappear. On September 16, 1962, Telemaco Xavier was taken away by an alien hunting party that added the unfortunate man to its collection of seventeen chickens, six pigs, and two cows.

Xavier was last seen walking home along a dark jungle trail after he had attended a soccer match in the village of Vila Conceicao in northern Brazil. A workman at a nearby rubber plantation told authorities that he had seen a glowing round object land in a clearing. Three beings got out of the fiery vehicle and grabbed a man who was walking along the trail.

Rio de Janeiro newspapers quoted authorities who stated that they had discovered "signs of a struggle where the worker said the fight had taken place." To the Brazilian newspapers, it seemed evident that "Mr. Telemaco Xavier was kidnapped by a flying disc."

Was the Brazilian added to a collection of Earth life that was to be scrutinized, evaluated, or dissected in some alien laboratory? Or are the crews of some UFOs carnivores, who see no reason to distinguish between the flesh of chicken, pig, cow, and *Homo sapiens*? Either question conjures up some decidedly unpleasant images.

CHAPTER THREE

Taken Aboard Alien Spacecraft

Charles B., of Miami, Florida, is haunted by the memory of lying on a table in a small, metal-like room in some kind of vehicle. The light in the room was radiant, soft. He felt sedated, calm, and trusting. Shadowy figures were moving around him, and he sensed that they were deciding on the best procedure to follow with him.

Charles is convinced that during that experience, which he believes was more than a vision or a dream, something was implanted in his brain.

"Ever since then, I feel heat, pain, and hear a crackling, popping sound in my head, like some integration process is trying to happen. My perceptions have changed dramatically. The intensities of those sensations have decreased quite a bit, almost as if the graft has taken."

Charles's girlfriend is very open to his experience, and he is able to discuss freely his feelings with her—about the blackouts, the amnesia. and the urgent aching feeling that he experiences inside. "I know what's happening," Charles says. "I'm being prepared for something. I feel so impatient. I just want to get on with it."

Charles isn't certain when the initial abduction happened to him. It may have been one night when he was driving home late from work. It may have been another evening when he stopped by the side of the road because he felt sleepy as he was driving home from having seen his girlfriend. But he recalls seeing not more than ten feet above him a circular vehicle with a square bottom and a distended portion covered with white lights.

He remembers feeling groggy and hazy as he gazed at it. Then an invisible beam of energy shot through him, and his entire body, especially his chest, vibrated and got very warm.

"Something exploded in my heart," Charles said. "I felt afraid, but I was determined to trust. The sensation, like a kind of electrocution, lasted quite a while, then stopped, and I went into a deep sleep."

But in spite of the deep sleep, certain impressions come back to haunt him. He has begun to recall images of entities working around him in that circular craft, trying to decide what procedure to follow with him.

Charles is convinced now that seeing UFOs or extraterrestrials is not necessary for the complete evolution of his thoughts and his awareness. He sees the physical experience as a crude level of blending and communicating. He feels that the actual process of communion is on a much subtler level.

"I know them deep inside. I feel them. I sense them. But I can't see them. The old patterns of 'show me, then I'll believe' need to go. I feel the need to broaden and to sharpen my deeper levels of perception."

EXAMINED BY "LITTLE GREEN MEN" IN POLAND

A Polish farmer, Jan Wolsky, stated that he was led into a spaceship by "little green men," given some kind of examination, and then released unharmed. According to a number of Polish medical doctors and psychiatrists, Wolsky, seventy-one at the time, was telling the truth as he could understand it.

The farmer's encounter occurred on May 10, 1978, as he was driving a horse-drawn cart near a forest near his village of Emilcin, about ninety miles southeast of Warsaw. Wolsky remembers seeing two little men up ahead by the side of the road. When he got closer, he could see that they were not ordinary men. Their faces were a gray-green color. They had large, slanted eyes, and their hands appeared to be webbed and of a greenish color. Each of the entities wore a little one-piece jumpsuit.

The beings leapt into his cart and motioned him to continue and then to turn off the road. Wolsky saw there in the clearing something as big as a bus, only "it was hovering in midair, moving gently up and down like a boat at sea, and humming gently."

Wolsky was led to the craft by the two small entities. A platform was let down. The three of them stepped onto it and were taken up into the spacecraft.

Wolsky remembers the interior of the craft as dark and empty. There were a large number of birds that had been gathered from the forest. They appeared to be paralyzed.

Two more aliens joined the couple who had led Wolsky into the craft and motioned for him to take off his clothes. The farmer did as was indicated, and then one of the little men came toward him with some sort of device that emitted a clicking sound. The entities moved around him with this instrument, indicating that he should raise his arms and stand sideways, as if they were taking photographs.

Wolsky put his clothes back on when it was signaled that he could do so. He stepped to the door, turned around, took his cap off, and said good-bye. The four entities bowed back and smiled at him.

Three witnesses from the nearby village corroborated Wolsky's story of a strange craft that had hovered in the area. They testified that they had seen "a flying ship" rise out of the forest and soar away. Upon investigating the site, the villagers said that they had found bird feathers scattered around the area.

Additional investigations by authorities revealed that "little green men" had been sighted in other parts of the region in subsequent months.

A VERY PAINFUL EXAMINATION

In *UFO Magazine*, volume 2, number 3 (1987), Jozaa Buist gave the details of her contactee experience, which began on August 23, 1975. That was the evening that she and a friend noticed eight objects moving across the South Pacific night sky, coming in from the northwest.

A few days later, a peculiar stranger, a woman named Valerie, who had pure white hair and a baby-soft complexion, approached Jozaa and claimed to know her very well. A series of very peculiar occurrences followed, somehow all interconnected with the ethereal Valerie.

Jozaa's first abduction experience occurred in July of 1980 at about 11:00 P.M. She and her husband—who is in the military—were in Hawaii. They had decided to retire for the evening, and Jozaa had gone upstairs to take a shower.

Afterward, she was cooling off by the bedroom window, which faced west. She looked out and noted a dark, orange object heading east. After about three minutes had passed, she began to feel very faint. She called to her husband, and he came running in, asking what was wrong. Jozaa remembers that sweat was pouring from her face.

Her husband took her to the Navy Regional Medical Center. The doctor reported that all her vital signs were normal and that he could see no reason for her faintness. He asked Jozaa if she were in pain, and she answered that she had experienced no pain or discomfort whatsoever.

About a week later, Jozaa was sitting on the edge of her bed, planning to read a magazine and fall asleep. The next thing she recalled was lying on a metallic table and looking up at three people staring down at her. They were dressed in doctor's uniforms with operating masks over their mouths. Their eyes were slits.

Jozaa next became aware of three metallic disks hovering over her body, making whirring sounds. Jozaa told the doctors that the whirring noise bothered her, and they said they would give her something so the sound wouldn't disturb her.

"I was given a shot in the crevice of my right arm," Jozaa said. "I seemed to calm down, but I was wondering what was in store for me. Then I saw them take a rather large hypodermic needle. I asked them what they were going to do and if it would hurt. They said it would, but only for a little while."

The next thing Jozaa knew, she was back in her bed. She had fallen asleep. It was about 5:30 A.M. She got out of bed, and as she began to stand up, she screamed out in pain. She was unable to stand.

The excruciating pain lasted throughout the course of the day, becoming progressively worse. Finally she decided she could endure the pain no longer, so she went this time to the Army Medical Center. The doctor there had seen her before for minor things, and he asked her what in the world was wrong with her, why she was crying so hard.

Jozaa really did not want to tell him. She felt she would be sent to see a psychologist, but against her better judgment she told the doctor the details of her bizarre memory of the operation the night before. Jozaa told him that she had never had any kind of surgery in her life.

The doctor looked at her stomach and asked her about the scar that was clearly visible. He also wondered about the mark on her arm. He left and came back with another doctor, and Jozaa had to tell her story all over again. To her utter amazement, the physician said that he had examined another patient earlier that same week who told a similar story.

Jozaa went to the lab and had blood work and other tests done. The results came back negative, as she had known they would. Her doctor told her that he couldn't prescribe medication, as the tests were negative. The pain continued for three weeks straight. She had a burning sensation in her navel, which she was convinced came from the needle that the aliens had inserted.

The next visit from the extraterrestrials was a lot more pleasant. Jozaa was offered a dark green "coffee." The entities showed her a map of all the military installations around the globe, which Jozaa did not understand at all. The entities also asked her why there was so much pollution on our planet. Didn't we understand that we were destroying ourselves? They asked her if she was afraid to discuss UFOs, and she replied that she was not.

Then Jozaa asked a question of her own: What was their purpose in entering our dimension?

They replied that they were simply observing us.

Jozaa and her husband were transferred from Hawaii to San Diego in 1982. She felt that she had left her agony behind in Hawaii, but that was not to be the case.

One night as she lay down on the couch to fall asleep, she was awakened about 7:30 P.M. with her arm hurting her severely. She called her mother, who was living across the highway, and told her that her arm was bothering her.

When Jozaa arrived at her parents' home, her mother asked her to remove her blouse. She turned Jozaa's left arm around and asked, "My God, what is that on your arm?" Jozaa's mother got a mirror so that she could see for herself.

"I stopped dead in my tracks," Jozaa stated. "I am dark complexioned, and this mark on my arm was pure white. It was the mark of a triangle. I measured it, and it was two and one-half inches on all three sides. A line stretched from the top of the triangle, and there was a white dot in its middle.

"My arm hurt badly. It went on for three weeks before I went to the Navy Regional Medical Center. The doctor said he could not X-ray it because it was so swollen, and he asked me about another lump or a bump that he felt. He gave me some medication."

Jozaa came back a month later, and the doctor who had treated her had been transferred. The new doctor seemed very understanding, and she had an X-ray taken. Eventually he called her name and said, "Let's see what the picture is going to tell us, if anything."

The doctor looked at the X-ray and explained that he could not make head nor tail of it. He decided to send it back to the hospital.

When Jozaa returned, the doctor began to act indifferently toward her. Jozaa wondered where her X-ray was and what the official medical interpretation of it was. The doctor replied that he had no answer for her, but it looked as though there were a disk in her arm. He would not discuss the matter any further.

Later Jozaa had to undergo a physical examination with another doctor for the adoption of her two daughters, and when the doctor looked at her arm, she asked about the strange marking. Jozaa was sent for X-rays; two days later, the technician who took the X-ray said, "I wouldn't touch your arm with a ten-foot pole."

When Jozaa, understandably disturbed, asked him what he had meant, he replied that he couldn't explain it to her.

After six months, Jozaa reports, the pain began to subside. She had finally given up on going to doctors. After all, she had received no tangible results. She had come to the conclusion that they didn't really understand what had happened to her.

In September of 1986, her husband was transferred to Nevada, and Jozaa thought she would see what could be done there. Once again she went to the Navy Regional Medical Center where they took more X-rays. She is still waiting for an answer to the mystery of the unexplained markings on her arm.

"I could ask the 'persons' who put them there, but who are they? I hoped that more could have been done by the naval doctors, but they could only go so far," Jozaa said. "Civilian doctors sort of shunned me. The marking is still visible to the naked eye. I have told my story over and over again, knowing that these extraterrestrials are real. I did not imagine them."

THE MULTIPLE EXPERIENCES OF ABDUCTEES

In the course of the numerous hypnotic regressions that he conducted with UFO abductees, Dr. James Harder said that he had found much evidence to support the theory that there is a means employed to find and to reexamine abductees at various intervals, sometimes throughout a person's lifetime and sometimes without them being aware of it. "It's as if some sort of extraterrestrial group of psychologists is making a study of humans," Dr. Harder observed.

Dr. Harder and others have discovered that a high percentage of people who have been abducted have undergone multiple experiences with UFO entities. Most abductees who have had more than one experience with UFO aliens usually undergo the first encounter between the ages of five and nine. These abductees remember the alien as friendly and quite human in appearance. Upon further hypnotic regression and careful probing, however, the investigators have learned that the entity did not look human at all.

In most cases, the entity usually tells the child that he will be back to see him throughout the course of his life. He also admonishes the child not to tell his parents about the encounter. In a great number of cases, memory of the encounter is somehow blocked out of the child's mind.

Harder has also discovered that during the adult abductee experience, those men and women undergoing the encounter will often report having a vague memory of their abductor, and they will say things such as "I feel I've seen this entity before."

In Dr. Harder's opinion, the multiple UFO abduction is not a random occurrence. "If it were random, the possibility of it happening to the same person more than once is extremely remote."

In 1981, Barbara Warmoth, of Franklin, Ohio, was abducted twice. According to this mother of six, she was studied by eerie seven-foot aliens with yellow, catlike eyes and pointed chins.

Her first experience, on February 15, occurred when Mrs. Warmoth saw an incredibly brilliant light filling her bedroom at 2:00 A.M. She got out of bed and saw through her window a saucer-shaped craft hovering nearby.

Under hypnosis, she learned that she had been aboard the UFO for one hour and fifteen minutes. The entities who examined her tried to reassure her. They told her that they had come from a planet called Antares and that they intended her no harm.

Her second abduction occurred on August 19, 1981, as she was driving along an interstate highway near her home. This time her experience occurred in broad daylight, and she lost two hours after the blinding glare of a silvery UFO forced her to pull off the road.

The being that examined her was dressed as before, but on this occasion, its face was uncovered. Mrs. Warmoth was now able to see more than just slanted yellow eyes. She could perceive that the being had no ears, a long, thin nose, pointed chin, and thin, almost colorless, lips.

Mrs. Warmoth was placed in a large chair in a room that seemed to her very much like a laboratory. She saw what appeared to be electronic equipment everywhere. The being gave her a glass of greenish liquid, and then the experience was over. She awakened back in her car, unharmed.

The Ohio housewife is convinced that the aliens will abduct her again, but she emphasizes that she has no fear of the beings. They have never mistreated her, and they tell her repeatedly that they intend no harm to come to her.

CARL HIGDON'S FANTASTIC JOURNEY

In 1976, a forty-one-year-old Wyoming oil-field worker claimed to have been kidnapped by alien beings while he was hunting elk in a remote wilderness area.

Carl Higdon of Rawlins, Wyoming, said that he was lifted aboard a spacecraft and taken millions of miles to another planet where he saw other earthlings living with alien beings. It was Higdon's impression that the aliens had been taking people from Earth for many years, as well as a sizable stock of various animals and fish. Higdon was given a physical examination, told that he was unsuitable for their needs, and returned to Earth.

Dr. Sprinkle, the University of Wyoming psychologist and UFO investigator who would regress Barbara Schutte to explore her 1981 close encounter, hypnotized Higdon a number of times and gained remarkable details of the experience.

Higdon had taken his rifle and a borrowed company truck to the north edge of the Medicine Bow National Forest to hunt wild game. About 4:00 P.M. he walked onto a rise and spotted five elk grazing in a clearing a few hundred yards away. He picked out the largest buck, lined it up in his telescopic sights, and pulled the trigger.

He could not believe his eyes when the powerful bullet from his magnum rifle left the barrel noiselessly and, in slow motion, floated like a butterfly for about fifty feet, then fell to the ground.

Higdon heard a twig snap, and he turned to face a strange-looking man who appeared humanoid but was like no human that Higdon had ever seen before. The entity was over six feet tall, about 180 pounds, and had a yellowish skin color. The being possessed a head and face that seemed to extend directly into its shoulders, with no visible chin or neck. The humanoid had no detectable ears, small eyes with no brows, and only a slit of a mouth. Higdon spotted coarse, golden, straw-like hair sticking out from the being's head. Two antenna-like appendages protruded from its skull.

The entity was wearing a tight-fitting, one-piece suit, similar to the outfits that scuba divers use. It also had a thick metal belt with a pointed star at the buckle. There was an unidentifiable emblem just below the star.

The alien being raised its hand in greeting to Higdon and floated a package of pills in the hunter's direction. Higdon remembers that he swallowed one of the pills upon the direction of the entity.

The next thing Higdon knew, he was inside a cube-shaped object with the being and at least one other alien, together with the five elk. Higdon was strapped to a seat with a football-like helmet on his head.

Then he underwent a bizarre trip through space in a small, transparent craft. Most of the details of Higdon's fantastic journey were gleaned during the hypnosis sessions with Dr. Sprinkle.

Higdon told the doctor that he witnessed portions of what appeared to be a futuristic city of tall spires and towers and revolving multicolored lights.

After his physical examination, he was returned to the space vehicle. When he looked out of the transparent sides, he observed five other beings who he felt were definitely humans. Higdon observed three adults and two children. The children were female, and one of the younger adults was female. In spite of his inquiries, Higdon was given no information about the earthlings.

The UFO set down again in Medicine Bow National Forest. Higdon was placed back in his truck without incident. His rifle was returned. He was relieved of the pills that the being had given him.

Although he was dazed by the strange experience, Higdon managed to radio for help. Then he apparently blacked out until he was found several hours later.

Higdon spent three days in the Memorial Hospital of Carbon County in Rawlins, undergoing extensive tests and rambling and shouting about four-day pills and men in black suits.

Higdon feels he has recovered from the experience and would like to forget about the entire incident, but he knows that somehow he will never be able to accomplish such a memory loss.

Some of the physical evidence that apparently supports Higdon's story is the fact that he was able to recover the spent seven-millimeter bullet that was crushed when it smashed into what may have been a force field between him and the elk that he intended to shoot. Forensic experts say that the odds against retrieving a spent bullet after it has been fired are normally millions to one.

The truck that Higdon was driving was found six miles from the location he last remembered being in it. It was mired in a sinkhole considered inaccessible to normal two-wheel-drive vehicles. Members of the search party looking for Higdon also observed strange lights and bizarre phenomena in the area.

Higdon apparently experienced a miraculous healing of a tubercular-type scar on his lung. A problem with kidney stones also disappeared after his trip to outer space. Dr. Sprinkle has observed that such unexplainable recoveries from ailments often occur among people who claim to have been examined by alien beings.

The question that remained in Dr. Sprinkle's mind was whether Higdon had actually had a physically real experience—whether he was actually taken aboard a craft and flown somewhere—or whether he was mentally programmed to *believe* that the trip to outer space had occurred.

"Anyone who has the technology that is apparently available in the area of UFO phenomena perhaps has even more sophisticated technology as far as consciousness alteration is concerned," Dr. Sprinkle said. In his vast UFO research, he said that he has come to believe that in some instances the aliens—whether they be extraterrestrial or possibly spiritual beings—may be making the abductees *think* that they are looking at a spacecraft.

Whoever the intelligences may be that are confounding us, Sprinkle has commented, it appears that they are trying to teach us that the world is more complex than we have formerly believed. "I think we are being taught to perceive that science and religion are not separated in this world—to recognize that as we gain greater technological knowledge, it will enable us also to gain spiritual awareness."

THE EERIE BRIAN SCOTT CASE

In February 1976, UFO researcher Timothy Green Beckley conducted an extensive series of interviews with the abductee Brian Scott. At that time Scott was a thirty-two-year-old draftsman for a Mission Bejo firm and the father of two, who stated that he had been repeatedly taken aboard a strange craft piloted by beings from an alien planet.

Scott's first abduction reportedly occurred in the Arizona desert near Phoenix in 1971, and he claimed that another had just occurred on December 22, 1975, in Garden Grove, California. In between, Scott said, there were three other terrifying sessions with the aliens and repeated visits to his home by balls of light and a transparent being that called itself the Host.

Scott believed that his involvement with the alien beings began on his sixteenth birthday, October 12, 1959. He had been coming home from celebrating when he observed a ball of light hovering over his dog. The ball was oval shaped, semisolid, becoming more solid toward the center. It was six to eight inches in diameter and reddish-orange.

The ball of light came right at his head until it was just a few inches from his face—then it shot straight up. Scott believed that, at that time, he had received some sort of communication from the ball through thoughts and pictures that were apparently transmitted directly into his mind.

It was more than twelve years later, on the evening of March 14, 1971, that Scott was transported aboard a hovering craft with a purple light emanating from its underside.

Scott had no knowledge of why he had chosen that particular evening to drive into the desert near the Superstition Mountains outside Phoenix. He remembers standing alone, seeing a strange craft fly overhead. Then he felt within a "pulsating, pulling feeling" that lifted him upward, into the vehicle.

Incredibly, Scott found that a friend of his was already inside the craft. The two of them were taken into a small room that began to be filled with a fog or a mist. Then they were confronted by four or five "very horrifying" creatures. Scott described them as having gray skin like that of a crocodile or a rhino, with a thicker patch of hide over the front torso.

Scott and his friend were disrobed and then led off in different directions. He was either carried or made to travel without bodily movement. The beings were seven feet tall, according to Scott, and looked like a combination of Earth animals. They had three fingers and a thumb kicked over to one side.

After undergoing a physical examination, he stated that his mind was transported to an alien world where he observed more of the strange creatures walking about a planet of jagged peaks in a misty atmosphere. After the mind trip, he was rejoined with his friend and returned to the ground. The last memory he had of the strange craft was a terrible odor, like "rotten socks, as if someone hadn't taken their shoes off for twenty years."

Scott's next experience also occurred in the desert near Phoenix on March 22, 1973. At that time he began to receive the impression that not only was he under observation by the beings, but that he was being slowly educated by them.

Resultant poltergeistic phenomena in the home is very often associated with UFO contactee or abductee experiences. Tim began by asking Scott about the kind of manifestations that had been occurring in his household.

> *Scott: There are streaks of light. A white light just streaks its way through the house, filters, and then just goes very quickly. Then there is the ball of light itself in the house and outside the house. There have been pure flashes, as if you put a flash cube right up to your eyeball. The light blinds you. You see it for just a few seconds, and then it goes. There is another object, a rather odd, brown-shaped thing that has from time to time shown up. It dashes around the room in crazy directions, and every time that it does, it creates some damage to the home. All the electricity and all the circuits in the house have melted, frozen, and burned up.*

> *Beckley: What happened on the day when your wife was sent to the hospital?*

> *Scott: She had been to work, pretty much handling everything that was going on around her. Then I got a call that she wasn't feeling very well. I brought her home, and after about fifteen minutes of sitting there talking with her, she was saying several things, none of which made any sense to me or to her.*

> *She said that she had been in the bathroom and suddenly felt hands all over her body. It was as if someone had broken into the house and molested her. When she calmed down and started making explanations to me about what the hell was wrong with her, it was as if, from her description, the guys I had seen aboard the craft in 1971 had visited her. This is odd, because she*

has never even seen any sketches that I made of those entities.

Beckley: *So this was an actual materialization—if you want to call it that—of the entities in the house?*

Scott: *I don't know what it was.*

Beckley: *But she was so upset that you decided to take her to the hospital?*

Scott: *Later that evening, it seemed as if she was okay. I was on the phone, and the baby was getting into everything so I couldn't carry on the conversation. I got up and went looking for my wife. I heard a bumping sound and a moan coming from the bathroom. My wife was on the floor, hyperventilating.*

I got her up and onto a chair in the living room. I was on my way to call her mother when she just fell flat on her face. I called the paramedics, and while they were on the way, she got up and fell down again. Then she began to become hysterical.

It took four paramedics to hold her down. She was throwing people around as if they were tissue paper. Guys were thrown backward against the furniture. Finally they loaded her up in the ambulance. I came back in the house, and the baby was not in the playpen. I panicked, because I couldn't find our one-year-old child. I ran back in the house. The dog was yipping at the back door. We finally found the baby sitting over in a corner of the patio. A one-year-old baby who got out of a playpen!

Tim Beckley asked Scott about the Host. "There is one entity that comes through that calls itself the Host, whatever that means," Scott attempted to explain.

"It speaks in what sounds like some kind of computerized language. The voice seems to come out of me, an inner voice that is not mine. The entity says that I am one with it. It says, 'I am; I am' or 'You are one with me.' When asked if it has a name, it will just come back and say, 'I am; I am.'

"The other night we heard some strange sounds coming from the bedroom. I began to speak in a foreign language that we later found out was Greek. Where that came from, I don't know. I wrote in Greek *backward*. On top of that, I was writing with my left hand, and I am right-handed.

"This voice was talking. We asked who it was, and the name Ashtar came out.

"Then it began to use the name Ashtar and speak to my wife. It told her things about her past that only she could know. This went on for a while, then it went on to say it would give her all the money in the world. It only wanted one thing in return—her soul."

Beckley pointed out that it sounded as though diabolical entities might be coming onto the scene, attracted by the extreme vibrations. He also observed that *Ashtar* sounded very much like *Ishtar*, an ancient Babylonian goddess.

The Host told Brian Scott that he would return on December 24 in the year 2011. He would descend on the spider figure in the Nazca lines. From there he would go to other ancient city sites where Scott and concerned parties were to construct pyramids.

Beckley interviewed J. D., an investigator associated with a civilian UFO investigation group. J. D. said that when he was first contacted by Brian Scott he thought the man totally out of his mind, but as he began to investigate, he became more curious and intrigued. He was especially impressed when the voice tapes that he had taken of various entities, which either spoke through Scott or from other areas around the house, appeared to produce prints different from the abductee's normal voice.

Beckley pursued this matter, learning that the mechanical voice of the Host "lacked all harmonics and seemed to be nothing but a series of small ripples."

Beckley knew that even if a person tried to disguise his voice or attempted to imitate another person's voice, the voiceprints would still reveal it as the voice of the deceiver. Each voice is very much like a fingerprint. There are individual characteristics in each voiceprint that designate a particular speaker. To learn that the voiceprint analyses of the various entities' voices were allegedly different fascinated the investigators.

Beckley asked J. D. how he would differentiate between what may have originally been an abduction case and the various types of poltergeistic phenomena that now seemed to prompt Scott's resultant trance state. Are they one and the same? Are they closely related mysteries? Or are they entirely different aspects of a more general phenomenon?

J. D. indicated that he was aware that there had been other cases such as Scott's. The manifestations of balls of light streaking through the homes of contactees and abductees apparently are more frequent than many investigators realize.

J. D. mentioned that one voice, a horrible voice, came through and claimed to be Beelzebub, the Devil. J. D. was convinced that the entity was simply trying to frighten away the investigators.

Beckley commented that the contents of the messages that Scott had relayed to him all seemed to be very sophomoric in content. Although a great deal of material was coming through from the alleged aliens, it did not, in Beckley's opinion, have any substantial value to its content.

Beckley spoke further with a technician who claimed to have analyzed the various voices connected and associated with the Scott case. He, too, indicated that they were quite different from one another. The company for whom the technician worked had wired Scott for twenty-four hours a day for one week. They used a four-channel recorder, recording different frequency spectrums on each channel. They recorded the vibrations of the house on the low-frequency channels, and the static electricity was recorded on the high-frequency channels.

The culmination of the project led them to conclude that Scott was not producing the various voices of his own will. Although the technician did not claim to be the final authority, he commented that some of the frequencies that they recorded were, in his opinion, so low that, generally speaking, a human voice could not produce them.

"Here again," he added, "we are going on the knowledge of standard speech, not necessarily something that is unusual in nature. But for all practical purposes, I am convinced that Scott was not doing this of his own will."

Beckley and other investigators formed a kind of consensus as to the basic material that formed the Brian Scott story. The entities that were contacting Brian Scott seemed to be of two basic groups.

The primary group appeared to be multidimensional in nature, indicating only that they were from a time beyond all time. Those people were tall and appeared human, but they often wore a bulky mass of loose gray skin as their "cloak of sorrow."

The secondary group was composed of beings who were small, with frail bodies, milky white skin, large bald heads, thin lips, and enormous eyes. It was stated that these beings had a common rapport with Earth beings and, in fact, were responsible for the genetic evolution of human life on this planet through sexual implantations over forty-five thousand years ago. Supposedly this group, perhaps from the sixth of seven planets around the star Epsilon Boötes, placed a satellite in orbit around our moon twelve thousand years ago. These beings may be considered negative by some, but in reality they are working hand in hand with a cosmic good to elevate humankind's consciousness. It appears to be the mission of the secondary group of entities to adjust the genetic structure of *Homo sapiens*. This adjustment will move humankind higher along the evolutionary scale.

The multidimensional beings, the taller, more humanlike entities from "time beyond all time," have the power to veto actions planned by those beings of the secondary world, but they will not interfere unless humankind reaches a certain maturity.

Scott was told that ten specific gifts would be given to humankind through his channeling. Some of these gifts would be spiritual in nature, but most would bring new technologies to humankind. His mission, however, was to begin to design a pyramid in Tiahuanaco, Bolivia, which was to be built before 2011 on the site of an existing inverted pyramid. He was also to tell civilization about the relationship between the world of humankind and the multidimensional world surrounding it. In addition, Scott was to design a transportation technology that would move matter through space. He was to master quantum displacement physics and begin to develop a mind transference machine to be utilized to unite all humans. Such a machine would help to develop a philosophy of cosmic brotherhood.

These tasks, of course, would seem impossible for a combination of Einstein and Superman, but they are typical of the type of grandiose missions assigned to so many contactees and abductees. Although the assignment may be impossible to complete, the process of the contactees striving may implant ideas that others may eventually be able to develop into workable technologies.

Scott's own feelings as to why he was selected for such a mission were understandably confused. He theorized that he may have simply been in the wrong place at the wrong time. Or he may have been a reincarnated pyramid designer, a reborn spaceman. Each group of investigators had a different set of theories to describe Scott, but it is known that he was "touched" by the Host at age sixteen and the Host seemed to be a spiritual guide to those who originated in the "time beyond all time."

As with so many contactees/abductees, Scott appears to have been changed by his experiences. His wife commented that his intelligence "skyrocketed" after his December 22, 1975, UFO contact. After that time, their relationship became strained because of Scott's increase in scientific knowledge. Allegedly, second- and third-degree burn scars were removed from his abdomen. According to his wife, his mood became serious and determined, whereas before the experiences, he had been "just an average guy, a lot of fun."

The Master Ashtar appears in much of UFO contactee literature. One cannot help noting the ancient origin of the name Ishtar, Ashtar, Asta, described always as a god of evil or negativity in the Bible. Whether the original Ashtar may have been an extraterrestrial commander whose motives were misunderstood by primitive earthlings or whether he may have been more indifferent to the needs of humankind than some of the other ETs allegedly walking around in ancient times can only be the subject for a great deal of speculation. Ashtar seems to belong more to the contactees than the abductees, but there are instances in which those who claim to have been forcefully taken aboard UFOs describe an interaction with beings who represent themselves as emissaries of "Ashtar's Grand Plan."

BETTY AND BARNEY HILL: THE INTERRUPTED JOURNEY

The case of Betty and Barney Hill has been covered in the press, numerous books, and a television movie starring Estelle Parsons and James Earl Jones. It would seem that even a person with only a cursory interest in UFO matters would be familiar with this case. As a brief memory jogger, however, I will outline the details of this prototypical "interrupted journey," which has culminated in a mind-boggling mystery concerning a star map.

Betty and Barney, a couple in their forties, were returning from a short Canadian vacation to their home in New Hampshire when they noticed a bright object in the night sky of September 19, 1961, over Franconia Notch in the White Mountains.

Barney stopped the car and used a pair of binoculars to get a better look at it. The light soon revealed itself to be a well-defined disklike object moving in an irregular pattern across the moonlit sky. Barney walked into a nearby field to get a better look. He saw the object plainly, and he was able to see what appeared to be windows—and from the windows, people looking back at him!

Barney was terrified as he got back into the car and raced down the road. Then, for some reason, he drove the car down a side road, where they found five humanoids standing in their path. Suddenly unable to control their movements, Betty and Barney Hill were taken to the UFO by the aliens.

The details of the Hills' story were elicited only under hypnosis, for they had a complete loss of memory concerning the nearly two hours following their initial contact with the humanoids. The Hills were returned, unharmed, to their car, at which time the beings told them they would forget the abduction. The UFO then rose into the air and disappeared from sight, and the Hills continued their journey home, oblivious to the whole event.

The remarkable encounter would probably have never been brought to light except for two factors: the inexplicable dreams they both had following the events aboard the UFO and the unaccountable two-hour gap in the journey home from Canada.

Barney, a mail carrier, and Betty, a social worker, continued to be puzzled. Finally Betty sought the help of a psychiatrist friend, who suggested that the memory would return eventually—in a few months. But the details of that "interruption" remained lost until they were revealed through the aid of hypnosis, conducted by Dr. Benjamin Simon, a Boston psychiatrist, in weekly sessions.

Under hypnosis the two revealed what allegedly happened that night. The individual stories of Betty and Barney agreed in most respects, although neither knew what the other had disclosed until later.

Both told of being well treated by aliens from space, much as humane scientists might treat laboratory animals.

They were then given a hypnotic suggestion that they would forget what had happened aboard the UFO. Their induced amnesia had apparently been broken only when they were rehypnotized.

The two hours aboard the craft consisted of various physical examinations, but the key to the whole event, and the factor that may be conclusive in giving the story credibility, is the star map that Betty claims she was shown while aboard the UFO.

Under hypnosis in 1964, Betty drew her impression of the map. Her map concurred with other, professionally drawn, star maps, which is in itself remarkable, since Betty had little understanding of astronomy. But there was a big bonus factor—her map showed the location of two stars called Zeta 1 and Zeta 2 Reticuli, allegedly the borne base of the space travelers. The existence of the two stars was not confirmed by astronomers until 1969—eight years after Betty "saw" the star map aboard a "spaceship." As an added zinger, the two fifth-magnitude stars are invisible to observers north of Mexico City's latitude.

The case of Betty and Barney Hill remains one of the most baffling and thoroughly documented of the abduction cases.

ABDUCTION BY MONSTERS AT PASCAGOULA

October is often a very slow month for UFO activity, but 1973 proved to be an exception to all rules, starting with the report of two Mississippi fishermen who told authorities that they had been taken aboard a flying saucer that looked like a giant fish.

Charles Hickson, forty-five, and his fishing companion, nineteen-year-old Calvin Parker, were fishing from an old pier in the Pascagoula River, near the city of the same name in Mississippi. The men reported seeing a fish-shaped object, emitting a bluish haze, approaching from the sky.

The craft landed, and the men allegedly were taken aboard by three weird creatures with wrinkled skin, crab-claw hands, and pointed ears. The men claimed to have been examined, then released.

Sheriff Fred Diamond of Pascagoula told investigators that the two men were scared to death when they reported to him and that he feared they might be on the verge of heart attacks.

Their story was interesting enough to draw the attention of Dr. J. Allen Hynek of Northwestern University in Chicago, who had served as scientific consultant to the air force's Project Blue Book, and Dr. James Harder of the University of California, who had the men hypnotized. They then revealed their traumatic experiences aboard the strange craft.

Harder commented, "These are not imbalanced people; they're not crackpots. There was definitely something here that was not terrestrial, not of the Earth."

"Where they are coming from and why they are here is a matter of conjecture, but the fact that they were here on this planet is beyond reasonable doubt," commented Hynek, who added: "The very terrifying experience of the two men indicates that a strange craft from another planet did land in Mississippi."

Hynek concluded that although the men could be hypnotized, their experience was so traumatic that it was necessary to progress slowly with them.

This is the story the two men told:

They were fishing for hardhead and croakers from an old pier near the Schaupeter Shipyard, a sun-bleached skeleton of a barge dry dock, at about eight o'clock on the evening of October 11. Suddenly a UFO hovered just above them. "There was me, with just a spinning reel, and Calvin went hysterical on me. You can't imagine how it was," said Charles Hickson.

According to the report of the sheriff's office at Pascagoula, Hickson related that the luminous, oblong craft landed near them. Three creatures paralyzed him, floated him to their craft, placed him in front of an instrument that resembled a big eye, then put him back on the pier.

Calvin Parker was not able to add much to the report. He apparently fainted when the creatures approached them, and he said he did not know what had happened inside the strange craft. After a couple of days, the two men refused further interviews with the press.

TERROR IN KENTUCKY

The night of January 6, 1976, will live long in the memories of three Kentucky women who were returning home from a late supper when they were abducted by a UFO crew and put through a torturous ordeal for more than one hour.

The three women, all reportedly of the highest moral character, were Elaine Thomas, forty-eight, Louise Smith, forty-four, and thirty-five-year-old Mona Stafford. All live in or near Liberty, Kentucky. Two of the women are grand-mothers, and Mrs. Stafford is the mother of a seventeen-year-old. None of the three could recall the full details of their experiences until they were placed under hypnosis by Dr. Sprinkle.

It was 11:30 P.M. as the three women drove toward their homes from Stanford, Kentucky. They were about a mile west of Stanford when they witnessed a large disk come into view.

"It was as big as a football field!" stated Mrs. Smith, who was driving the car that night. She continued her description by stating that it was metallic gray, with a glowing white dome, a row of red lights around the middle, and three or four yellow lights underneath.

The UFO first stopped ahead of them, then cir-cled around behind their car, at which point the car sud-denly accelerated to eighty-five miles an hour. The others screamed to slow down, but Mrs. Smith found that she had no control over the car. Some force then began dragging the car backward. At that point the three women lost consciousness and remained unconscious for the next eighty minutes. The events that allegedly took place were brought out later under hypnosis.

The three women remembered vividly what had taken place during the lost eighty minutes—they were brought aboard the UFO to undergo complete physical examinations.

Elaine Thomas reported that she had been lying on her back in a long, narrow, incubator-like chamber. The humanoids looked to her like small, dark figures, which she estimated to be about four feet tall. She reported that a blunt instrument was pressed hard against her chest, causing much pain, while something circled her throat.

Each time she tried to speak, she was choked. She cried softly under hypnosis as though reliving a horrible ordeal. It felt like hands pressing on her throat, and she could see shadowy figures passing around her. "They won't let me breathe—I can't get away!" she cried.

Under hypnosis Mrs. Smith said that she had been in a dark, hot place, and that something had been fitted over her face. She begged the occupants to let her see, but when they did, she immediately closed her eyes, as whatever she saw was quite frightening. She could not describe the beings, however.

"Help me, Lord, please!" she cried. She told investigators that the interior of the UFO was very dark and that she was quite frightened. She pleaded with the humanoids to let her go, to let go of her arm.

She finally cried out, "I'm so weak I want to die!" Still later she asked them if she could go, and the next memory she had was that of seeing a streetlight.

Mona Stafford's memory was of lying on a bed in what seemed to be an operating room, with her right arm pinned down by some invisible force while three or four figures dressed in white gowns sat around her bed.

Apparently Mrs. Stafford was not as overcome as the others, but she did say that she seemed to remember being tortured, and that her eyes felt as though they were being jerked out of her head at one point. At another time, her stomach felt as though it had been blown up like a balloon. Next she reported that the humanoids were pulling at her feet, then bending them backward and twisting them. "I can't take no more!" she screamed, then lapsed into silence.

The next thing the three frightened women could remember was driving to Louise's home. They should have arrived about midnight, but they noticed the time was actually 1:30 A.M.—nearly one hour and twenty minutes were missing from their lives that night.

Louise reported that her neck hurt. When Mona examined it, she saw a strange red mark like a burn that had not blistered, about three inches long and an inch wide. Elaine's neck had the same type of mark on it.

The frightened women called a neighbor who lived next door to Louise, Lowell Lee. After hearing what they could recall of their adventure, he had the three women go into separate rooms and draw what they felt the strange UFO looked like. The three drawings looked very much alike.

Although the burn marks were gone in about two days, the three women still could not account for the time loss, nor could they recall anything from the time the car was pulled backward until they were driving near Huston, eight miles from where they first saw the UFO.

Following the hypnosis sessions, they were given polygraph tests by Detective James Young of the Lexington police department. Young, in a sworn statement, said, "It is my opinion that these women actually believe they did experience an encounter."

Dr. Sprinkle stated that the three women, in his opinion, had specific impressions that they had been observed and handled by strange beings. He felt it would have been impossible for them to fake their reactions, and he commented that their experience during the time loss was similar to reports provided by other UFO percipients.

Sheriff Bill Norris, of Lincoln County, Kentucky, said that there had been a number of UFO sightings in the county that January.

In an article by Bob Pratt that appeared in the *National Enquirer* on October 10, 1976, Len Stringfield, a director of Mutual UFO Network, who investigated the whole incident, commented, "This is one of the most convincing cases on record."

The report of abduction, the memory loss, the missing time, and the shape of the UFO are all familiar to UFO investigators. How truly bizarre it is that such an incident can become commonplace to those who research such matters. With professional detachment, the case is filed together with others of its type. And yet for those men and women who undergo the trauma and confusion of a UFO abduction, it becomes a memory of terror that may never fade.

CHAPTER FOUR

Teleportations and Materializations

On June 3, 1968, *La Razon* carried the story of a physician and his wife who claimed that a mysterious fog had transported them from the province of Buenos Aires to Mexico.

Dr. and Mrs. Vidal had attended a family reunion at the home of Señor Rapallini in Chascomus, a town situated near National Route 2. They arrived in the late hours of the afternoon, ate an elaborate dinner, then began the drive back to their residence at Maipu, a city in Buenos Aires province. Another couple, near neighbors in Maipu and also relatives attending the same celebration, left at the same time as the Vidals. Both cars set out on Route 2 just a few minutes before midnight.

The first couple, whose names were not released, arrived at their home without incident, but because they had previously agreed on it, they awaited the arrival of the Vidals before they retired for the evening. After waiting for several minutes, the neighbors began to fear that the doctor and his wife might have met with an accident. They decided to retrace the route back to Chascomus to look for the Vidals.

Strangely enough, they drove the eighty miles back to Chascomus without seeing a trace of the Vidals.

Once again they decided to drive back to Maipu on Route 2. This time they studied every foot of the highway. They now had to accept the grim possibility that the Vidals might have overturned in the steep banks, and because of the night shadows, the wreckage had gone unnoticed on the first trip. Again, nothing. A visit to the Maipu hospital gained only the information that no accidents had been reported that night.

The Vidals had vanished without a trace—or so it seemed until nearly forty-eight hours later when the Rapallinis' telephone rang.

"Do not worry about us," Dr. Vidal said, attempting to calm his friends.

"But where are you?" Señor Rapallini wanted to know. "Where are you calling from?"

"I am calling from the Argentine consulate in Mexico City," Dr. Vidal said.

There was an incredulous gasp and an exclamation from Señor Rapallini.

"I am sorry, but I cannot give you more details at this time," Dr. Vidal apologized, "but we will be flying back to Buenos Aires. Here, I will give you the date and hour of our arrival so that you might meet us at the airport in Ezeiza."

An astonished group of friends and relatives gathered to meet the plane from Mexico City at the appointed hour. Dr. Vidal still wore the same clothes that he had been wearing on the evening of their disappearance. Mrs. Vidal, the victim of a "violent nervous crisis," was taken directly from the airport to a private clinic.

Although Dr. Vidal had been warned by the consulate not to issue any public statements about his strange experience, he did relate enough details of their "interrupted journey" for La Razon to piece together the story from bits and snatches of interviews with the Vidals' friends and relatives.

According to Dr. Vidal, he and his wife had left the city of Chascomus a few minutes before midnight and were traveling normally on Route 2. They had been listening to the radio, and Dr. Vidal stated that he had been driving at a speed that would enable him to keep the taillights of his friend's automobile always in view.

Then, shortly outside the suburbs of Chascomus, Dr. Vidal found their car enveloped in a dense fogbank. He slowed down.

He was aware only of blackness. No more sensory impressions.

Suddenly it was bright daylight. Dr. Vidal blinked his eyes, looked around him. He was on a strange and unfamiliar road. His wife lay sleeping beside him. He roused her, then got out to inspect his automobile.

He was startled to discover that every bit of paint had been scorched off the car's surface, as if someone had burned it off with a blowtorch.

"But where are we?" his wife asked, trying hard not to give way to tears.

"All I can say is that we are on the side of a road," said Dr. Vidal, shrugging. "How we got here and what happened to our automobile I cannot say."

Dr. Vidal got back inside the auto and was pleased to discover that the engine still worked perfectly. "At least the scoundrels did not burn up our motor," he said.

"Look!" Mrs. Vidal gasped. "There's another motorist. Wave him down and ask him where we are."

Dr. Vidal acted upon his wife's suggestion. The motorist's answer to his question did nothing at all to settle their nerves. "He says that we are outside of Mexico City!" Dr. Vidal told his startled wife.

"H-he jokes with us," Mrs. Vidal protested.

But after several motorists had been stopped and the identical reply had been given to the same question, the Vidals were forced to accept the fact that somehow, in a manner far beyond their understanding, their return trip to Maipu had been detoured to Mexico City, another continent and several thousand miles away from their destination. Later, a calendar told them that their catnap during the trip had lasted forty-eight hours.

The Vidals were taken to the Argentine consulate in Mexico City, where they gave a full report of their bizarre experience. There they made the telephone call to the Rapallini family in Maipu, Argentina. Their automobile was removed to an American laboratory for examination, and arrangements were made for the Vidals' return to Argentina.

La Razon commented: "In spite of the halo of fantasy that the story of the Vidals seems to wear, there are certain details which do not cease to preoccupy even the most unbelieving: The entrance of Vidal's wife into a Buenos Aires clinic; the proved arrival of the couple on an airplane that arrived nonstop from Mexico; the disappearance of the car; the intervention of the consulate; the serious attitude of the police in Maipu in regard to the event; and the telephone call from Mexico to the Rapallini family—which was confirmed by *La Razon*—make all of this acquire the status of a matter worthy of being considered in these times of space adventures and fantastic appearances of flying saucers."

For a later edition, a reporter from *La Razon* asked Professor Alejandro Eru, secretary of the Argentine College of Parapsychology, for his views on the Vidals' strange, befogged aerial trip. Professor Eru responded by telling the journalist of three similar cases of mysterious transportations.

According to the parapsychologist, who is a professor of humanities at the University of La Plata, a man who lived near Bahia Blanca suffered a dizziness when a strange aerial craft appeared before him. Ten minutes later, he came to, but he found himself in Salta, one of the northernmost provinces in Argentina. The police in both locales communicated with each other immediately, and the man's automobile was found in Bahia Blanca in precisely the spot where he claimed it would be from Salta.

The second case related by Professor Eru happened to a professor of law on the faculty at Santos. He claimed that a flying disk sucked him aboard and took him on a remarkable aerial tour before releasing him.

"The third case," said Professor Eru, "involves a most widely known and highly responsible painter, sculptor, and theatrical artist, whose initials are B. S. P. He was for many years the director of the Art Salon of the Municipal Bank."

B. S. P. testified that he had been detained by a blond, Nordic-appearing man, "with eyes so clear that he seemed blind." The blond man spoke in a guttural, unintelligible language, with "friendly mannerisms." B. S. P. got a glimpse of the stranger's flying disk just before a wave of dizziness engulfed him.

"When he awakened," Professor Eru said, "he saw that he was flying along with three other beings. One of them, very gentle, interrogated him in a language also unintelligible, but our compatriot understood, or at least believed he perceived the man's thoughts, by telepathic communication."

The UFO occupant told B. S. P. not to be frightened; they would return him to Earth at exactly the same spot from which they had plucked him. In a few minutes, B. S. P., who said that he was in a kind of swoon, claimed that he saw the terrain of Japan, France, and later, Chile. When he awakened from his "trance," he was standing in precisely the spot at which he had first encountered the strange blond man.

"In none of these cases do the witnesses speak of hallucinations," Professor Eru told La Razon.

Nor do they speak of having been intoxicated or doped. What caused these phenomena? Well, unfortunately,

we Earth people cannot answer that conclusively. . . .
Our supposed friends from other worlds . . . possess,
without a doubt, some type of electromagnetic wave
with which they attract any nonmagnetic object from
the surface of the earth so that they may use this
object in their studies.

How does parapsychology view this? Well, from this
angle, we find it noteworthy that these beings can
manage telepathy with such great mastery because
they seem to be able to make our minds understand
things which cannot be expressed by speech. At
the same time, they capture, without difficulty, our
answers. Nothing more, with any seriousness, can be
said of these phenomena at this time. But for some
special reason, the Americans have kept the Vidals' car
in their powerful laboratories to examine it!

BEAMS OF LIGHT AND DARK FOGS

In his article "UFOs and Light," in *UFO Report* (April/June 1976), Charles Bowen tells of Swedish engineer Sven Sture Seder, who was driving his Volvo on a highway near Ojebyn on September 20, 1971. Suddenly a black object raced past and ahead of Seder, and in a few minutes, he found himself driving into a cluster of light beams that he said descended from a light in the sky. When his car was surrounded by these beams, he felt "an unnatural force from behind," and although the Swedish engineer slammed his foot on the brake, his car continued to roll forward until it entered a dense, floating mass of dark smoke.

Seder said later that the smoke or mist was so dense that even the rays of light were absorbed by it, along with his headlight beams. As quickly as he entered the smoke, he seemed to leave it, and he found his car had gone a considerable distance. As he recovered his composure, he could see a dark, kite-shaped object speeding away ahead of him.

Bowen suggests that either something had gone wrong with the exercise or whatever intelligence was controlling the attempted abduction had discovered that Seder was unsuitable material for its purpose.

On May 30 to 31, 1974, a young Rhodesian couple, Peter and his wife Frances, who requested that their last names be withheld, were driving toward the frontier post at Beit Bridge, Rhodesia, and were about six miles past Umvuma when they had their encounter with the unknown.

Peter was driving their Peugeot 404 at more than sixty miles per hour when his wife saw what she thought was a policeman dressed in a metallic-looking suit. The policeman appeared to be standing beside the road with a walkie-talkie radio. At first they suspected a speed trap, but then they spotted a UFO, a luminous object with a beamlike spotlight that revolved like a lighthouse beacon. This light was bluish and switched on and off regularly.

While Frances kept an eye on the UFO, Peter watched the road and listened to a radio station broadcasting from Lourenço Marques in Mozambique. Suddenly, they both remembered, the interior of the car became very cold, and the couple wrapped themselves in blankets.

Peter, however, now had another problem to worry about, as the Peugeot was traveling at ninety-five miles per hour—and to his great dismay, he had no control over it. He slammed hard on the brakes, but that had no effect on the vehicle, and the steering wheel appeared to be locked in position. The car accelerated to more than one hundred miles per hour, and to make matters even more frightening, the headlights went out.

Somehow they were able to pull into the filling station at Fort Victoria and stop. The station attendant, who was dressed only in an undershirt and shorts, looked at Frances in astonishment when he saw that the woman was wrapped in a blanket, shaking with cold.

At 5:30 A.M. they were on their way again. Once again, Peter was in control, and it seemed as though the nightmare, whatever its cause, was over.

Six miles away from the filling station, Frances saw that the UFO had returned to the same position as before, high to the left of the car, while a second one had appeared and was directly over their automobile.

In amazement, Bowen writes, the couple now found themselves driving down a perfectly straight road (where it should have been twisting), with grass, bushes, and swamps on both sides of the path (where there should have been dusty, dry terrain). Their radio still blared out, the speed of the car had now risen to well beyond its capability—more than one hundred and fifteen miles per hour—and Peter had no control of the vehicle at all.

In spite of the enormous stress and tension of such a bizarre situation, Frances found herself falling asleep at about 6:15 A.M. during a peculiar, dull, overcast, gray dawn. Peter drove on, but he had lost all trace of time, and he said later that he felt as though he were in a coma. Frances remembers awakening at about 7:00 A.M. when they were at the border customs post. The Rhodesian officials laughed at the couple who looked as though they had just arrived from the North Pole.

Since they had now driven one hundred and seventy-nine miles since leaving Fort Victoria, they decided that they would fill up their gasoline tank on the South African side where it was much cheaper. They noticed, however, that the odometer had recorded only eleven miles—and when they tried to fill up their gas tank it could only take twenty-two cents' worth.

South African UFO investigator Carlo Van Vlierden arranged with a doctor to put the couple under regressive hypnosis. Van Vlierden reported some remarkable facts that emerged from the interview while Peter was regressed.

Beams of light, Peter said, gave the car a complete mind of its own, and it seemed as though the car were telling him what to do—smoke a cigarette, light the lighter, switch the radio to another station.

The craft above the car, Peter reported under hypnosis, was sending down pulses all the time. He remembered trying to fight back, but he had been helpless.

A simulated screen was projected in front of the windshield and along the side windows, and it was on this screen that Frances and Peter saw the lush vegetation.

"We traveled the whole way," Peter said, "completely above the road."

During the regressive hypnosis experience, Peter remembered being put back down on the road about two miles before reaching the border customs post.

Frances, Peter said, had been put to sleep by the radio, which carried the voice of "them." An entity was beamed down into the backseat next to his sleeping wife, and the being announced to Peter that it could assume any form that Peter might wish to see.

Peter was told that they had determined that he was a deep-trance subject, and it was obvious to the investigators that he had fought a mental struggle against them when they tried to erase his memory. As the hypnotic sessions continued, Peter said that he could see inside the craft via the beam of light. All the entities looked the same, but they could assume any form.

The aliens were apparently friendly to Earth people, but they revealed that they could not make contact openly because the majority of humans would not understand them. They were mortal, they suffered death, and they traveled by time, not by light.

They had mastered every Earth language, and Peter said it was revealed to him that these beings will change Earth eventually by slowly introducing their manner of doing things.

One of the final revelations that Peter uttered while he was regressed was to state that there are now thousands of these beings, these UFOnauts, among us here on Earth. They are posing as clerks, bus drivers, typists, businessmen, students, teachers. They make contact only when they want to. They never do anything directly. They are manipulators and schemers.

Although that final statement might cause some people great alarm, Bowen comments that there is no way of knowing that it was not a phony story, planted deliberately in Peter's mind as part of a deception to conceal some other purpose.

There is no question that a lot of the information that the beings put into Peter's mind has a familiar ring to it. It seems very much the same type of information relayed to so many other contactees. If Peter's story that aliens by the thousands walk among us, posing as ordinary people, is true, then the great question remains, *Why?* And what is the true purpose of such an extensive program of infiltration?

A DEMATERIALIZATION FROM A HOSPITAL BED

Some years ago, I recorded a most remarkable account of a man who had either discovered a doorway to other dimensions of reality, along with an ability to dematerialize his physical body, or been granted these unique talents through his interaction with UFO entities. As I stated in *Mysteries of Time and Space*, if the following account were not attested to by a very matter-of-fact young man currently associated with one of the largest, most prestigious hospitals in the Midwest, I would be extremely hesitant about sharing it with the public.

I made contact with the man I'll call William through a correspondent who had taken a course in medicine with him. According to my correspondent, William had not mentioned his experience during the several weeks' duration of the course, but one day after class, he discussed it over a cup of coffee. According to William, the following occurrence took place in a hospital in Hawaii in 1968. William was then about nineteen years old, serving in the medical corps, and assigned to the military section of the hospital.

Briefly, this is what happened:

A bedridden patient, in traction and totally unable to move, with pins through his tibiae and femurs, told William that he would be gone that night for one hour to join his friends in a UFO. He said that William might accompany him *if* he truly believed in UFOs. William indulgently told the patient that he would be unable to join him that night, as he would be busy.

Later, during bed check, true to his word, the patient had disappeared, leaving the metal pins on the bed. An extensive search of the hospital and the surrounding grounds by military policemen failed to produce any trace of the supposedly immobile man.

The patient, William said, was about sixty, and a veteran of World War II. He was kind of a sixty-year-old hippie. He had been on an LSD trip when he walked in front of a tractor-trailer and broke both of his legs in several places.

He had a Spanish-sounding kind of name, something like "Espinia." He had bushy eyebrows, shoulder-length blondish hair, very large eyes. He had a round face, a flattened nose. He was about five foot six and a bit chubby.

"Espinia was always discussing his weird techniques for meditation," William remembered, "and he had a strange accent. By the time I was assigned to that hospital, I had already been around the world a couple of times, and I'm a bug on accents anyway, but I simply could not place Espinia's.

"Espinia was always talking about peace, love, brotherhood. You know, how we should get out of the war in Vietnam.

"The night he disappeared, I was working the eleven [P.M.] to seven [A.M.] shift," William continued. "When I made the initial bed check, Espinia told me that he would be gone for about an hour, and he reminded me that I could come along if I wanted to. I chuckled and walked on to see about the rest of the patients.

"Espinia was in a six-man room, but that night he was alone. My post was almost across from his room. When I sat at my desk, I could survey the entire corridor. No one could get on or off the floor without my seeing them. And, of course, there were nurses, doctors, interns, and MPs walking around.

"When I checked Espinia's room a bit later—maybe out of curiosity—he was gone. The traction weights were hanging there; the pins were on the bed; but Espinia was gone.

"I put out an alarm, and MPs and other hospital personnel searched the place thoroughly. But Espinia was gone. Some patients said that they had seen a bright light, a very bright light, on that side of the building, and that would have been just before Espinia's disappearance."

William said that a man would faint from the pain if he tried to pull the pins out by himself. "This guy was lying in that bed with both legs up, his femurs broken. Think of the terrible pain of trying to crawl under such conditions. It would be impossible!"

But when hospital personnel and MPs next checked the disappearing patient's room, he was found to be back in traction, pins in place. The patient had been gone for one hour. He told his interrogators that he had been with his "friends."

"After searching that hospital—and even the grounds—for an hour," William told me, "somebody looked in Espinia's room, and there he was again, back in traction. Everything in its place. A doctor on the floor said that while it might be possible for a man to pull the pins *out*, it would be impossible for anyone to shove them *back in* by himself.

"Four MPs grilled him for hours, but Espinia wouldn't even reply to their questions. When they finally left him, he looked at me and told me that I could have come along with him, but his UFO friends knew that I didn't really believe in them. He said that he and his friends had spent a delightful hour flying over the Hawaiian Islands and chatting about metaphysics.

"When I bawled him out for having caused such a disturbance in the hospital, Espinia became a bit sheepish and said that the next time he went flying with his friends, he would leave his body there and just go with them in his mind."

TELEPORTED ABOARD ALIEN SPACECRAFT

Psychic-sensitive Clarisa Bernhardt has established a record of accurate predictions, especially in regard to forecasting earthquakes. Internationally known as a seer, Clarisa has also undergone a series of encounters with aliens.

During interviews with Clarisa, my associate Hayden Hewes and I learned that she had experienced a mental projection to a craft before she underwent a physical teleportation.

Clarisa Bernhardt: On two or three occasions they have "transported" me while my body was still on the bed. They took my consciousness aboard the spaceship. I felt very humble about getting to go there, so I just listened to them. At the time, they gave me some predictions, but they felt since I was sensitive toward earthquakes, the scientists and so forth would be more impressed if I gave them the two largest earthquakes of that year.

There were three aliens, and they were about five eight to five ten in height. They had on silver-colored uniforms. They had helmets on, so I wasn't able to see the details of their faces. But I did not feel that their heads were out of proportion with their bodies. In other words, they did not present a grotesque appearance as far as our human understandings are concerned. When I asked them who they were, they told me that I might think of them as Space Brothers. A lot of their people have been in suspended animation, and they said that our people on this particular planet will learn more about that in the near future. They gave me the name of their world, but it was unpronounceable as far as I was concerned. They said that they had come from several galaxies away. Sometimes they travel back and forth in time, as well as in space.

They told me that they were very concerned about some of the things that have been happening here on Earth. They're afraid that we are going to have problems with nuclear power. If we blow ourselves up, it's going to mess up some things in the universe.

They are here on a peaceful mission—a mission of peace and love. They would like to communicate with everyone. There will be more contact, but for the most part, they feel that man has only evolved above certain primitive emotional characteristics.

Can you recall the physical description of the interior of the craft?

Clarisa: One particular area to which they took me was as big as a city block. They took me into an area where crew members were sleeping. There were rows of six as far back as I could see.

Did they speak verbally or mentally in the communication area?

Clarisa: They spoke to me mentally, because each time they contacted me there was this light that came forth. It was as if I were hearing them this way.

Was the sound a mechanical one, a "human" sound, a computerlike sound? Did you hear it in your voice or their voice?

Clarisa: It was in English, and it was not a computerlike sound. There was strength behind it. There was no doubt when they told me something.

What do you think their purpose is in giving you the information about the earthquakes?

Clarisa: They said they wanted to assist me. They knew that I was trying to help people. Although I have this ability through my own development (they said my metaphysical studies had made me more sensitive),

I was a clear receiver when they had a reason to contact me.

Did they give any indication as to how long they had been aware of life on Earth? Do you feel their presence is a recent thing or that they have been around for a long time?

Clarisa: From what I understood, it's as if they have been waiting for the people on Earth to grow up. This is the best way I know how to express it.

I do definitely feel from my contact with them that they are going to come here and intercede. I don't mean they are going to come in and start fighting with people. But because of their abilities, they have a lot of knowledge that can help us.

Do they make reference to a Supreme God force throughout the universe?

Clarisa: They did not make any reference to this. I let them talk to me, and I asked a minimum of questions. I feel that they are a higher and more evolved form of life than we are at this point. I'm sure there are more things in the universe to behold than our present consciousness can understand.

Did they give any indication as to any other projects that they might be encouraging?

Clarisa: The main area in which they want to help our Earth is in the advancement of scientific avenues. If we can prove ourselves worthy, they would like very much to have us in a position whereby we might communicate and learn a whole lot about the universe.

Following our interview, Dr. Dean Sterling, a hypnologist and hypnophysiologist, hypnotized Clarisa and taped the following testimony regarding her teleportation aboard an alien spacecraft:

Dr. Sterling: What happened on your way to San Jose? What was the first thing that happened that was strange to you after you left your house that day?

Clarisa: I am late. I now see there's not much traffic. I'm not going to be late. But I feel funny. I feel dizzy.

There's no reason for it. There's a strange feeling in my forehead. It's coming all over me. I feel awful. I feel like I'm falling out of my body. I have to drive the car! I hope I don't faint.

How fast are you driving?

Clarisa: I'm driving fifty-five. I'm going to slow down, because I'm concerned I might have an accident or go off the road. But there is something that I can't explain. It's like it's in my forehead. Hurts my head. Something's happening. Voices are saying something.

What are the voices saying?

Clarisa [Beginning to get upset]: They're telling me, "Don't be afraid." I don't know if I'm dying or what. Wait . . . Wait . . . But they're telling me I'm not to talk.

I hear a man's voice. He says his name is Marisha. He says, "Don't be afraid." But I am afraid. [Crying] I cannot explain what's happening. I'm not driving my car. It's like there's a cloud over the car. I think I'm gonna faint.

What type of cloud?

Clarisa: It's like a mist. It's like a big thing coming around the car. I don't know what it is. Good God . . . They said not to be afraid. Wait! What's happening now? There's five . . . figures in front of me.

What do you see around you?

Clarisa: I can't . . . I am hearing something. "Don't be afraid." There's one coming forward. He's not speaking. But it's coming into my mind. He says his name is Marisha. "We are from another time and from another light. You will understand that we came not to hurt you. We must explain things to you."

Explain your surroundings. Can you see what's around you?

Clarisa: It's like we're in a building. It's circular. It's like we're in a circle, and there's a lot of equipment in here. My car is here. It's like we're in a big garage, except there's no window. It's like it's a very sterile place.

What are they telling you?

Clarisa: They're telling me that I will think of them as Space Brothers. They come only to communicate with me. They're sorry if they frightened me. I feel better. I feel that there's not as much to be afraid of now. I think I'm understanding now. Two of them are coming forward.

There are some lights around the room. Looks like a lot of instruments. I think it's similar to the panel of an airplane . . . the controls. A computer. There's lots of information, and there are things happening on it.

What are the Space Brothers saying?

Clarisa: They are telling me that I will learn more about them. They're sorry they had to frighten me, but it was necessary to show me that they had control. I must never doubt that they have control over me or anyone else. They do not ordinarily do this, but they said that it was important that I understand. They will be in contact with me in a few days.

Why are they in control of you? Why have they chosen to be in control of you?

Clarisa: They tell me that the reason they are in control of me is that they can control anyone. This is not their way of getting people to do things, but they wanted to impress me with the fact that it can be done. They say to me that because of the position I am in because of my radio show, they feel I can do much to further their mission.

They come in peace. They do not want to hurt anyone. But they do worry about us.

They say that humankind is on the verge of making a mess out of things. If we wish to destroy ourselves it is fine, but there is a responsibility to the universe. For this reason I can be of help to them by making others more aware that there are others in this universe besides those of us here on this planet. They say that we have not been responsible and that we have much growing to do.

They are not caught in time, as we are caught in time. We will learn of this in the future.

They told me that they would let me return and that I would then be conscious of my experience. They told me that they would contact me more in the future. And that I would not be frightened of them.

They told me that I would have confusion, but it would clear up. I must be impressed with the fact that they can control many people, but they do not wish to do this. I am concerned that they do not have feelings of compassion.

Why are you concerned that they don't have compassion?

Clarisa: They are good, but they are so cold . . . I have a feeling that . . . they have a different understanding of life from ours. Perhaps it is because they have so much knowledge.

There is a strange glow in here, and I think I'm told that I'm going back. They said they send only thoughts of love.

I'm starting to get dizzy again . . . but I'm not going to be frightened this time. I do believe them. There is a glow. It's like a fire. Then there's this mist, like a cloud, that's happening again.

They said that I may not remember for a little bit, but I will remember later on in the day. I will be contacted again.

I'm getting dizzy, and my ears are ringing. There's this hurting again. And I feel like my head's gonna pop. But I'm not going to be frightened. I just breathe deeper.

What's happening? I must have fainted! I don't know where I am.

Where are you?

Clarisa: This is weird. In San Jose!

Where in San Jose?

Clarisa: I don't know where I am. I have to drive down here to the road and look at the sign because . . .

San Jose is back the other way. It says Oakland is ahead. My God . . . How did I get here?

CHAPTER FIVE

An Extraordinary Interest in Human Sexuality

Dagmar and Carl R. have a farm in northeast Iowa about forty miles from the Mississippi River. One night in August of 1982, Carl observed what he called at the time a "lantern in the sky" that hovered over him while he was working late in the field.

Two days later, while he was getting the cattle home in the evening to milk, he noticed a strange, circular burn mark in the pasture. He paced the mark and found that it was roughly thirty feet in diameter and formed almost a perfect circle.

He called Dagmar out to observe the mark, and together they tried to determine what might have caused such a strange blight on their pasture. They investigated to see if it could be some kind of ant or burrowing insect that had eaten away the grass and made the ground appear brown and lifeless. They found no insect of any kind.

Dagmar suggested that if there had been any rain or thunderstorms it could have been a strike of lightning. But both of them agreed that even if there had been an electrical storm it was unlikely that a lightning strike would leave a perfectly circular mark.

The next morning, Carl noticed two more burn marks in the pasture, again perfectly circular, roughly thirty feet in diameter. He mentioned the new circles to Dagmar that night as they ate dinner, and they wondered what could possibly be causing such peculiar and unprecedented marks in their pasture.

The next evening, when Dagmar was bringing home the cattle, she saw what she thought at first to be one of the neighbors' children watching her from the cornfield. As she looked closer, she saw that it was not a child, but a strange-looking man, roughly five feet tall with an unusually round head and very large eyes that seemed to stare at her in an eerie, expressionless way. When she waved and called at the stranger, the smallish man raised his hand as if in answer to her salutation. Then he walked into the cornfield away from her.

Carl became quite upset with her report that night over dinner, because he felt that one of the neighbor boys might be spying on his wife. Dagmar assured him that it was not one of the children. They were on familiar and friendly terms with each of their neighbors, and she would surely have recognized one of the young boys.

Carl became even more concerned at that point and suggested that Dagmar might have seen a runaway or lost child, who was somehow surviving in the cornfields or in the barn and outbuildings. That morning he said he would see if he could find any sign that someone had been living on their farmland.

Although he did not spend a great deal of time in the search, he was able to examine areas where it seemed logical to him that someone might be trying to survive, and he found no trace that anyone had attempted to do so.

In October of that year, while Carl was working late in the field preparing for the annual corn harvest, he was startled to see the glowing "lantern" return to the sky above him. It appeared to be the same object that he had seen in August.

Although he tried to remain oblivious to the object, it seemed to be hovering above him, even following him up and down the corn rows. He became nervous and disconcerted and went back to the farmhouse, where he asked Dagmar to come out and witness the strange object.

Dagmar was able to see the object, too, and they stood and watched it for several minutes before it suddenly moved high into the night sky and then sped off at a great rate of speed in a westerly direction.

About three the next morning, Carl was awakened by the sound of cattle bellowing nervously in the stockyard. As he got out of bed and looked out the bedroom window, he saw a disk-shaped object hovering above the barnyard. It was glowing a kind of greenish color.

Then, on the ground, he saw shadowy figures, smallish men moving about. He was about to awaken Dagmar when he turned suddenly to see three of the entities standing in the doorway of their bedroom.

He shouted in alarm, and Dagmar awoke. She let out a short scream when she saw the entities.

According to both Carl and Dagmar, after their initial outbursts of fear, they were both overcome with a marvelous feeling of peace and tranquility. They permitted the entities to take them by the hand, to lead them down the stairs, out through the front porch, across the yard and walk them toward the disk-shaped craft that was now hovering over the orchard near the farmhouse.

Dagmar remembers clearly that, as they approached the craft, a stairway seemed to move down noiselessly and touch the ground. She recalls walking side by side with her husband up the stairs and into the dimly lit interior of the hovering craft.

At that time, although she did not feel greatly fearful, the experience became less pleasant for her. She felt very apprehensive when the entities asked her telepathically to lie down on what she described as "a large piece of smooth white ivory." As she lay there, manacles moved out to clamp themselves around her wrists and ankles. When she found she was unable to move, she became quite alarmed.

At that point, an entity, who she somehow felt was more female than the ones she had seen previously, moved toward her and began to remove her nightgown. The nightgown was cut away so that it could be slipped over her shoulders. The entity that seemed more feminine was, in Dagmar's recollection, a bit shorter than the other entities, with even larger eyes and somehow a bit more pointed nose and fuller lips. Other entities, who stood around and watched the procedure, appeared taller, with only nostril openings rather than a pronounced nose, and with tight, expressionless lips.

The feminine entity who removed Dagmar's robe made a musical, humming sound as she extended her hand. One of the attending entities gave her what appeared to be a syringe.

Dagmar remembers shrinking from the pointed object and watching it enter the flesh of her abdomen. But she felt no pain, and all the time, the entities surrounding her were making soft, musical sounds that she described as almost like the cooing of doves.

Dagmar recalled that even though she was confident that she would not be harmed and that the entities continually sought to reassure her of that fact, she still felt a natural kind of apprehension as to what would happen to her. Would she be cut open like some livestock animal being slaughtered? Or would she be deemed healthy and therefore eligible to be taken away from Earth to some faraway and unknown planet?

The entities around her moved swiftly. She remembers a bit of her hair being snipped, one of her fingernails being clipped, and portions of her skin being rubbed with what she said felt like her husband's after-shave—a liquid that left a kind of cool, tingling sensation. She was also aware of the odor of her own burning flesh when the entities removed a mole under her left arm.

Carl's recollection is of being taken into a small cubicle where two entities approached him with a green vial filled with a liquid that appeared to be swirling of its own volition in its container. Carl remembers that the entities did not form human words when they communicated with one another, but that they made a humming sound that reminded him of birds cooing. The words of communication seemed to form in his own mind telepathically, and he knew clearly that they wished him to drink the greenish-colored liquid.

Carl felt no apprehension in doing this because he was convinced that he would not be harmed by the entities. When he drank the greenish liquid, however, he briefly had a terrible feeling that he might have been poisoned or drugged. He began to feel very hot all over and then chilled, as if he were getting a fever.

To his embarrassment he could feel that he was uncontrollably attaining an erection. To his further embarrassment, this seemed to be what the entities were seeking to achieve. With a pleased kind of humming noise, they moved aside the drawstrings of his pajamas and clamped some kind of mechanical device over his erect penis.

Within a few moments Carl had achieved orgasm, and he was very much aware that the semen was being drawn from his body and deposited into a vial. Although the beings remained as expressionless as they had with Dagmar, Carl knew that they were pleased with the specimen that he had produced for them, and two of the entities left the room in a very excited state. Carl explained that he was aware of their pleasure by the fact that their cooing noises had become very high-pitched and staccato.

Carl and Dagmar awakened in their bedroom the next morning. They were concerned at first because they had overslept and fieldwork was in full swing for harvest time. They did not remember the experience for several days.

When they began tentatively to discuss the mysterious dreams they were having, dreams of being taken aboard a craft and given a physical examination, they became very uneasy. Dagmar even became nauseated with the discussion, and they dropped the entire matter until around Christmastime, when Carl began to have nightmares.

What seemed to trouble Carl the most was the memory of his semen being taken from him. A religious man, very orthodox in his beliefs, Carl began to have nightmares of his semen being used to produce alien babies on some other world.

When he at last shared this fear with his wife, Dagmar began to express more elements of the experience and her own conviction that it had not been a dream. As they began to discuss the matter freely, they recalled the scorched marks that they had found in their pasture, and Dagmar remembered vividly the strange man she had seen watching her from the cornfield. Upon further discussion, Carl remembered watching the entities in the barnyard apparently collecting a wide array of specimens for some unknown purpose.

While the young Iowa couple can remember no further UFO interaction since that particular autumn, they both admit to being nervous about having another encounter. Carl, especially, feels that he was used. Dagmar speculates that bits of her skin and tissue might have been removed in the examination, and although she does not claim to be an expert in such matters, she wonders if enough of her body could be cloned in a way to interact with whatever embryo or fetus might have been fathered by the semen that was taken from her husband.

Not wanting to sound like victims of some science fiction thriller, the young couple have theorized that they might have been used in some strange program to create hybrid beings. Perhaps, they suggest, Carl's semen was used to impregnate an alien female or an Earth female who is somehow influenced by and under the control of alien beings. In either event, they are uncomfortable with the experience and with the memory of the encounter. Both of them feel as though they may have been used in ways opposed to their normal expression of will.

Dagmar has gone even further in her speculations by suggesting that if bits of her body could have been used to create a clone, and if Carl's semen could somehow be used at a future time to impregnate such a clone, then alien beings could be breeding their own brand of humans as part of an organized program to create an army of humanlike robots that would be totally under control of aliens in their master plan to conquer Earth.

A PREGNANT WOMAN IS EXAMINED

UFO investigator Richard Siefried was told by Pam Owens that she was taken aboard a UFO on November 25, 1978, while she was expecting a child. She was nineteen at the time, and she had no memory of the abduction until she was hypnotically regressed. Then she was able to give full and fascinating details of her encounter.

Mrs. Owens told Siefried that she was paralyzed and able to move only her eyes. She lay helpless on a table and stared up in terror at two weird-looking creatures.

According to Mrs. Owens, their heads were hairless, oversize domes, their eyes big and sunk back in their skulls. The greenish skin covering their bodies was coarse. Each hand had four fingers that she described as being twice as long as a human's. And to her terror, one of those strange hands was holding a long silver needle, preparing to plunge it into her stomach.

Pam Owens pieced the story together after a series of hypnotic regressions. She had been standing with her husband Chris and their twenty-month-old son Brian, gazing in absolute amazement at an object that they understood to be an approaching spaceship. Then suddenly she was inside of it, alone and terrified.

At that time, Chris had been stationed with the US Army in West Germany. They had been visiting friends near Trier, and it was on their way home that they saw the UFO. It was shiny, metal, oval, and perhaps one hundred feet long. Pam always remembers a red blinking light under the craft.

Pam told her husband to get out of there, to drive away fast, to get away from the UFO. But when they got home, they discovered that it had taken them two hours to drive what should have been a half-hour journey.

Later, when they returned to the United States, Pam contacted a UFO research organization, and the investigators there suggested that she undergo hypnotic regression by a psychologist. Under hypnosis the young mother's mind revealed what had really happened that night.

When the UFO first appeared, they had driven off the main road and stopped in a clearing. She and Chris got out of the car and held Brian while they stood there waiting, unafraid.

The next thing she knew, she was lying on a table. She could move only her eyes. She was in a small room filled with yellowish-white light. The ceiling blended into the walls, so there were no corners.

The frightened young mother asked where her little boy was. She recalls a flat voice answering, "We are taking care of him."

Then she asked about her husband and begged the entities not to hurt him. The voice kept repeating that everyone was safe.

After she had calmed down, two of the entities moved into her line of vision. She remembers that they looked almost like mummies. They had very tiny noses and straight lines for mouths.

One began to speak to her, reassuring her that everything would be all right, but his mouth didn't move, and it seemed to Pam as if he were somehow talking through his eyes.

Suddenly the entities pulled up her shirt, exposing her five-months-pregnant abdominal region. Their hands touched her, and she began feeling dizzy and sick. She began to cry for the safety of her unborn child, but the entities continued to examine her, as if they were checking on the baby's size and the way it was lying in her womb.

Then she saw the needle, and she became truly frightened, crying out, "No, no, don't hurt my baby!" But one of them stuck the needle right below Pam's navel—in the exact spot where she later found a puzzling pimple.

She remembers that the needle hurt a great deal, and she felt as though she were going to vomit.

The next thing she knew, she was standing by their car again, holding Brian in her arms as before. The family was watching as the spaceship rose into the air.

Pam feels that the ninety minutes missing from their lives have not been explained. In her opinion, the creatures, wherever they came from, were basically friendly and were examining a human body to see what it was like.

"Four months after my experience, my daughter Kelli was born—perfectly normal," Pam Owens said. "Brian, who couldn't talk when it all happened, doesn't seem to know what occurred that night."

SEXUAL INTERCOURSE ABOARD A UFO

Although there are many reports of people in the United States who claim to have undergone "close encounters of the most intimate kind," one of the earliest accounts of human-alien sexual intercourse came from Brazil.

British researcher Gordon Creighton translated the accounts of this incident, which were originally published in the Brazilian magazine *O Cruzeiro*. Shortly after the publication of Creighton's translations in *Flying Saucer Review* (1965), the author received correspondence from Dr. Olavo Fontes, one of the original investigators of the alleged act of procreation between beings from two different worlds. Dr. Fontes included transcripts of the initial declaration made by the young man on February 22, 1958, and the official report on his medical examination.

Dr. Fontes stressed that although intelligent, the young man, Antonio Villas Boas, had little formal education. Such matters as UFOs and alien beings simply were not within the interest range of Antonio and his fellow farmers near the town of São Francisco de Sales in the state of Minas Gerais.

According to his own deposition, Antonio first saw the UFO through his bedroom window one night after his family had had a party on their farm. Antonio described it as being "like the light of a car lamp shining downward." In the darkness, Antonio and his brother Joao watched the "light penetrating through the slats of the shutters, moving toward the roof, then shining down between the tiles."

About nine days later, the strange light reconnoitered Antonio a second time while he was plowing a field with the family tractor. Again his brother also witnessed the light "so bright it hurt the eyes."

On the next night, October 15, 1957, Antonio was plowing alone when an egg-shaped object came at him and began to hover above his tractor. The twenty-three-year-old farmer realized that escape was impossible on his slow-moving tractor and that the soft earth turned up by his plow blades would impede escape on foot.

"I could see the shape of the machine clearly," Antonio said in his deposition. "It was like a large elongated egg with three metal spurs in front. On the upper part of the machine there was something that was revolving at great speed and also giving off a powerful fluorescent reddish light."

When the object began to land, Antonio observed three metal supports being lowered to take the weight of the craft on the soil. The young farmer admitted that he lost what little self-control he had so far preserved. He only managed to run a few steps, however, before someone grabbed his arms.

He wrenched himself free of the grasp of his first pursuer, but he soon found himself being boxed in by three other "men" who grabbed his arms and legs and lifted him off the ground. Antonio, a well-muscled Portuguese-Amerindian, said that his abductors were about his height (about five foot four) and strength. Later, in his deposition, he stated that he thought he could have given a good account of himself on a man-to-man basis.

As the kidnappers carried Antonio toward the egg-shaped craft, the young farmer began to scream for help and to curse the strange men. "My speech seemed to arouse their surprise or curiosity, for they stopped and peered attentively at my face every time I spoke."

Once inside the machine, Antonio stood in a brightly lit room as two of the men held his arms and others gathered around to talk about their catch.

"I say 'talked' only as a way of putting it," Antonio told Dr. Fontes and the other men who recorded his deposition. "For in truth what I was hearing bore no resemblance whatever to human speech. It was a series of barks, slightly resembling the sounds made by a dog."

After the aliens had finished "discussing" the situation, Antonio was stripped naked. The husky young farmer tried once again to resist such manhandling, but the aliens seemed to try to make him understand that "they were a polite people."

Deciding that it would be simpler to comply with their wishes, Antonio allowed himself to be thoroughly examined. A chalice-shaped glass flask, with a nozzled tube, was applied to his chin, and some minor operation was performed that left a scar still visible to Dr. Fontes and the investigators. Another tube was applied to the young man's side, and Antonio saw his blood "slowly entering the chalice until it was half full. Then I was bled once again on the chin, from the other side, where you gentlemen can see this other dark mark like the first one. This time the chalice was filled to the brim and then the cupping-glass was withdrawn. The skin was grazed at this place, too, burning and itching, just as on the first side."

When the aliens had finished with their respective tasks of pricking and poking poor Antonio, he was left alone to rest on a couch. He had not lain there long before he became aware of a gray smoke that began to enter the room from some tubes protruding from the walls. The smoke had a suffocating odor "like painted cloth burning," and Antonio gave vent to his nausea by vomiting in a corner of the room.

After a few more minutes, Antonio seemed to adjust to the nauseating odor and began to breathe easier. It was then that the startled young farmer had a most surprising visitor. The door to the room was opened, and a well-proportioned and totally naked woman joined him on the couch. In spite of the bloodletting and skin sampling he had just endured, and in spite of his embarrassment at finding himself naked in the presence of a woman who might be from outer space, Antonio felt himself responding to her frank advances.

Later Antonio told Dr. Fontes that the aliens must have doused him with an aphrodisiac to have made him enter into such a rapid sexual union with the woman. He described the woman as having had large blue eyes that seemed to slant outward, a straight nose, high cheekbones, a nearly lipless mouth, and a sharply pointed chin.

After the sexual act had been completed, one of the alien men appeared in the room and barked to the woman. Before she left the room, she turned to Antonio and pointed to her stomach and to the sky.

The man handed Antonio his clothing and indicated that he should get dressed. It was obvious that the young farmer had served the purpose for which he had been obtained, and the occupants of the UFO no longer had need of his blood or his body.

If the aliens left satisfied that they had gotten what they had desired from their visit, it developed that they were not really as considerate of Antonio as they had seemed. The next day the farmer became ill. His eyes began to burn and a series of sores broke out on his arms and legs. In the middle of each of the sores was a little lump or spot that was very itchy. Two weeks later, Antonio's face became speckled with yellowish spots. These symptoms strongly suggest radiation poisoning or exposure to radiation.

During the period of his ordeal, Antonio had had plenty of time to study his captors while they had been busy examining him. It seems worthwhile to quote some of these observations because they bear great similarity to the descriptions of aliens given by so many who have witnessed UFO occupants. They are doubly valuable because this unsophisticated farmer from the primitive Brazilian interior had not previously been exposed to UFO contactee stories. If the language seems a bit too polished and literary to be that of such a man as we have portrayed Antonio Villas Boas to be, keep in mind that I am quoting from British writer Gordon Creighton's translation in *Flying Saucer Review*.

> [The aliens] were dressed in tight-fitting overalls made of a thick but soft cloth, grey in color, with black bands here and there. This garment went right up to the neck, where it joined a sort of helmet . . . of the same color, which seemed stiffer and was reinforced at the back by strips of thin metal, one of them being triangular and on a level with the nose. . . .
>
> . . . The height of their helmets must have corresponded to double the size of a normal head. It is probable that there was something else as well in the helmets . . . but on the top, from the center of the

head, three round silvery tubes emerged which were
a little thinner than a garden-hose pipe. These tubes,
one in the center and one on each side, were smooth,
and they ran backward and downward, curving in
toward the ribs.

The aliens wore five-fingered gloves of a thick material,
but that did not hinder them from gripping Antonio
tightly, nor from deftly handling the rubber tubes they
used for extracting his blood.

All of the members of the crew wore, at chest level,
a sort of round shield, the size of a slice of pine-
apple, which from time to time gave off luminous
reflections. . . .

The trousers were also tight-fitting over the seat,
thighs, and legs. . . . There was no clear separation
at the ankle between trousers and shoes; they were a
continuation of each other. . . . The soles of their feet,
however, had a detail different from ours. They were
very thick, two or three inches thick, and quite turned
up (or arched up) in front. . . . Despite this, the men's
gait was quite free and easy, and they were nimble in
their movements.

Gordon Creighton, a former British consulate officer
in Brazil, subsequently offered a few theories of the Antonio Villas Boas incident, labeling it "The Most Amazing
Case of All."

Antonio makes it clear that all of his little men were
wearing helmets with pipes coming from a device
located on their backs. The girl, who did not leave
the UFO, wore no such helmet or device, presumably
because she was "at home" in her own atmosphere. It
is of course true that Antonio had no helmet or breathing device either, and he claims to have been able to
survive in there. Let us not forget, however, that he
did have an attack of violent vomiting. . . . Does this
perhaps mean that their atmosphere, although dis-
agreeable to us, can nevertheless be tolerated by us,
and is not fatal? That, by contrast with this, our atmosphere is impossible for them? In such a case, would

not the obvious solution be to breed a mixed race, a
new race which would have inherited some of our char-
acteristics, including our ability to live in a mixture of
eighty percent nitrogen and twenty percent oxygen? A
new race, in brief, which is destined to live here, and to
populate the vast uninhabited areas of Brazil?

Have the UFO crews been seeding a new race, not only in the remote areas of Brazil, but also in such highly pop-ulated regions as California? I have on file the claim of a young California woman—that she was raped by an occu-pant of a UFO. There is also a deposition by her doctor who testifies to having treated the young woman for the prema-ture delivery of a stillborn baby that seemed to have been the product of highly dubious mixed breeding. Adequate documentation is not yet available on this case to estab-lish whether the girl has told the truth about her cosmic rapist or whether she has invented a highly original story to account for an unwanted pregnancy.

The fact that the woman's highly distraught condition was responsible for her miscarriage points up a very real factor that UFO breeders might have considered in their plan of mixing races. Not many earthwomen would have the emotional strength to undergo the trauma of rape by an alien creature and then calmly nurture its seed within their swelling wombs. Then, too, control of the pregnant mother would be extremely difficult while operating from a spacecraft. To bring her aboard for her nine-month ges-tation period would accomplish no more than a speedy mental collapse. If such a master plan of genetic manipula-tion is in progress, it would be much easier to deal only with embryo-stage humans taken from earthwomen.

Or, in the case of Antonio Villas Boas, bring an alien female, whose egg is ready to be fertilized, to the earth-man. The nature of male sexual response being what it is, the aliens need only provide a woman of their species who would most nearly correspond to an earthman's ideals of feminine beauty and douse the man with a powerful aph-rodisiac to help him overcome any innate shyness and fear. Presto, the whole problem of interbreeding becomes much easier to accomplish. The alien woman waits out her gesta-tion period in comfort, tended to by her own kind, and the earthman is left with either a memory lapse or a story that no one will believe. The aliens would gain another healthy product of crossbreeding; the inhabitants of Earth would receive only another case study to confuse their already ten-uous and hesitant efforts to derive a meaning from the UFO abduction reports.

Such cases as Villas Boas's one-on-one sexual encounter with an alien female are not nearly so common as those in which human males become unwilling semen donors.

"Don't think I'm some kind of nut," the letter began in a familiar plea.

"I never did believe in UFOs, but here is my experience. I will not reveal my name, but I live in Las Vegas. My hobby is women, and since I am married, I must be very careful.

"A girl and I were parked on a very lonely desert road. We had a blanket on the ground and were very busy in a certain act, when a very, very hot wave fell on us.

"I looked up to see two men, both about five feet, six inches tall, standing beside us in a soft light. They had on some kind of coveralls that looked like divers' suits. Their faces didn't look strange, but they had no hair on their heads. They spoke a language I couldn't understand. Behind them, hovering about twenty feet off the ground, was a craft that had a circle of small lights around its middle.

"The two men raised us up by our arms, and they felt all over our naked bodies. They pushed the backs of our knees and made us kneel, and one of them cut off some of my girl's hair, put it in a container, and pointed toward the sky. Then they walked beneath the ship, stood in the circle of a spotlight, and they were gone. The UFO disappeared from sight in ten seconds."

Another percipient reported a similar lover's-lane experience in which two UFOnauts allegedly lifted him off his girlfriend, milked his penis of semen, then hurried back to their hovering craft with their prize of human sperm sealed in a metallic container.

UFOS AND WOMEN

I first began receiving letters revealing the UFOnauts' extraordinary interest in our planet's females in 1966. During the years 1967 to 1970, I was inundated by reports from college women who claimed that they were being sexually troubled by UFO entities.

The letters all had a monotonous sameness about them, yet they contained a strong note of urgency, and they came from all sections of the country and from overseas.

"Dear Mr. Steiger," a typical letter began,

"I am not a nut; I am on the Dean's List at college, majoring in physics. This is my real name, and if you are suspicious of me, you can check me out.

"Last summer I saw a flying saucer at close range. It hovered over my car for several miles as I drove to my parents' farm home. It was definitely a metallic object.

"Shortly after that sighting, I was aware of something in my bedroom one night as I was preparing for bed. I could see nothing, but I could not shake a feeling of uneasiness . . . I was not yet asleep when I felt a pressure on the bed beside me. When I sat up, I saw nothing, but I felt *something* fondling my breasts. I wanted to scream, to get out of bed, but I was unable to move. . . . I remember nothing more until I awakened the next morning, but I have reason to believe that something made love to me while I slept. I believe this incident was associated in some way with my sighting the UFO."

A series of letters from a college girl in a northwestern state detailed how an invisible *something* began to annoy her in her bedroom after her sighting of a UFO. She described fainting spells and inertia that would come over her whenever she attempted to tell anyone of her experiences. She said she had fainted twice while writing to me, because *they* were interfering, censoring her thoughts.

In one of her letters she wrote,

"I felt something heavy at the foot of my bed. It moved to the far left. Then I felt nothing.

"A few seconds later I felt a warm radiation just to the left of where I was lying, so I moved to the far right of the bed. The heat began to radiate warmer and warmer, until I was sweating. I threw off the covers until it cooled off. Then after I was lying down again, it started to get warmer. This continued all through the night."

This young woman, who planned a career in law enforcement, continued to write, describing continued materializations and testifying that "aliens" walked among us, infiltrating our society on all levels.

"You can tell them by their eyes," she wrote, just before I received a farewell letter saying that "they" no longer wished her to write to me. "I have come to realize that they are here to help us, and we should cooperate with them."

An extremely lengthy letter from a young chemist told of her sexual liaison with an invisible UFOnaut after she had had a close sighting of an unidentified flying object.

"I lay on my bed one night, just dozing off. Then I heard the steady tread of footsteps coming up the stairs. I knew the doors were locked and that I was alone.

"I lay there in fear, as the footsteps came closer and closer. At last they stopped beside my bed, and as the bedclothes were torn off my body, I wanted to scream, but could not. I lay unable to move as the thing lifted my nightgown and mounted me. I knew, as only a woman can know, that the thing was male."

In *UFO Warning*, New Zealand author John Stuart told of a sexual assault on his research assistant that came after the two of them had seen a grotesque, misshapen entity while investigating a UFO report. Stuart stated that he felt the young woman should have stayed with him and his wife, but she had insisted that she would be all right in her own home. Barbara, the young woman, had been incorrect.

She immediately noted a peculiar odor the moment she stepped into her room. She undressed, bathed, but could not free her mind of the impression that unseen eyes followed her every move. Then, as she crushed out a cigarette and turned to put on her pajamas, an invisible *something* touched her on the shoulder. She found herself unable to move. The horror had begun.

For two and a half hours, an unseen entity had its way with her body.

"I concluded that the thing had been solid, even if invisible," Barbara told Stuart later. "There was, of course, no way of knowing exactly what it was like, and I tried to form a picture in my mind to fit it, but I gave that up in fear. I got into bed and eventually fell into a deep sleep, filled with nightmares. With the light of day, I again looked at my body and shuddered when I saw the scratches. It had really happened after all."

Barbara's flesh had been left covered with scratches from its contact with the rough-skinned space-age incubus. Her ribs also bore two brown marks about the size of an American dime. Barbara told Stuart that the thing had seemed only "clinically interested" in her body, as if it were engaging in sexual intercourse with an earthwoman more out of curiosity than sensuality.

Quite understandably, Barbara lost her enthusiasm for UFO research after her terrifying otherworldly rape, and she moved to another city. A few weeks later, John Stuart himself came face to face with "it"—a humanoid of grayish color.

The creature communicated with Stuart telepathically and told him that thirteen of them had encircled Barbara

THE UFO ABDUCTION BOOK

on the night that she had been attacked, but only three of them had taken part in her sexual violation.

"Why did you scratch her?" Stuart asked angrily.

"It was something we couldn't avoid," the entity replied.

"Where did the two brown marks come from?" Stuart demanded.

"They are there to remind her of us," the thing responded.

Psychic sensitive Ron Warmoth told me of a client of his who had also been given "marks" by which to remember her alleged extraterrestrial lover.

"I just looked at this woman in amazement when she told me that she was having sexual intercourse with a spaceman," Warmoth said. "She told me how she had seen a UFO and had established mental contact with one of the crewmen on board. Later, according to her, the spaceman had begun to materialize in her bedroom at night and make love to her."

The woman told the psychic that her otherworldly lover was well versed in sexual techniques, but that his torrid embrace often left her with round burn marks. Warmoth expressed some skepticism about the woman's story; then, before he could protest, the woman raised her skirt and displayed the evidence on her body.

"I can't say, of course, how the marks got there," Warmoth told me, "but her inner thighs and her stomach were covered with small round burns. It almost looked as if someone had placed a hot metal grid work against her flesh."

John A. Keel, author of *Strange Creatures from Time and Space*, did articles for two popular magazines on certain aspects of the bedroom invaders and was amazed by the amount of mail that those pieces drew. "Many readers wrote to tell us, sometimes in absorbing detail, of their own experiences with this uncanny phenomenon," Keel said. "In most cases these experiences were not repetitive. They happened only once and were not accompanied by any other manifestations. In several cases the witnesses experienced total paralysis of the body. The witness awoke but was unable to move a muscle while the apparition was present. . . . Such visions could possibly be created by some kind of hypnotic process or by waves of electromagnetic energy that beam thought and impressions directly to the brain. This would mean that the experience was not entirely subjective but was caused by some inexplicable outside influence."

Keel told me: "It is my conclusion that cases of UFO sexual liaisons are actually a mere variation on the incubus phenomenon. Induced hallucinations seem to play a major role in these cases. There may be considerable validity to the theory you expounded in one of your books [Demon Lovers] . . . that semen is extracted from human males in some succubi events and that this same semen is then introduced into human females in incubi incidents. The true nature and purpose of this operation is completely concealed behind a screen of deliberately deceptive induced hallucinations. Early fairy lore is filled with identical cases, as you know. And such sexual manipulations are an integral part of witchcraft lore."

WHY ALIENS ARE ATTRACTED TO EARTHWOMEN

Willard L. Wannall served in the military in Hawaii. When he was still in the service, he had a number of UFO experiences and began seriously to pursue the matter of flying saucers. Wannall claims that his superiors ordered him to cease all studies, but whether this particular research began to prepare him for his later contacts and abductee experience is a matter of speculation. At any rate, among the accounts of Wannall's interaction with the space beings is a very interesting passage that the entities gave him on the importance of earthwomen, and it seems appropriate at this point to quote some of that material from Wannall's privately printed book, Wheels Within Wheels and Points Beyond.

The responsibility now for the degree of evolutionary success each civilization achieves has been assigned to woman. She is supplied with a greater degree of the negative effect of light energy. Man is supplied with the positive effect.

There are several fundamental reasons for this. Primarily, the physical planes represent light manifesting at a lower frequency than the positive plane. As part of the cosmic plan to continue the process of creation-causing imbalances, each one of the two planes is influenced by a certain amount of opposite force existing in the other. This holds true for male and female physical bodies.

The strongest manifestation of the neutral phase of the cosmic root force's light energy, and the one that underlies all creation, is known as the love vibration. The female physical vehicle through its high negative charge, so to speak, attracts the love vibration more strongly than does the male vehicle with its positive charge. When a critical peak is reached, the love vibration being drawn by both male and female, even though not balanced, sets up an energy emotion that is a composite of the predominant force in the male and female blended with the neutralizing effects of the love vibration. In a sense, two such individuals are vibrating on the same frequency and are said to be in love with each other. Physical love can be considered as the drawing together on the physical plane of the positive and negative forces through the common denominator of the physical and emotional energy of the neutral love vibration.

All men and women are affected by each other through these unknown polarity changes, often to the extent that a deep love is manifested for one's fellow humans. Whatever the relationship, the progression of the entire cycle is controlled mainly by the degree of love for humanity possessed by the women. This is so because of the great concentration of negative forces composing her physical vehicle.

The Infinite One—a perfectly balanced neutral being—is a symbol of pure love, and contained within that love is the secret of all creation. The love vibration flows through the neutral phase of light energy from the very center of infinity—the seed atom of the omniverse—and becomes the strongest power in existence, endlessly creative in an ever-expanding cosmos.

The female physical vehicle was ideally wrought to provide the love vibration with a channel and a means through which its purpose of procreation could be accomplished. Just as love is the highest expression of light, so is woman potentially the truest exemplification and exponent of love on the physical plane. She possesses a far more complex emotional nature than man, as well as more fully developed intuitive powers, which increase her sensitivity to the higher vibrational levels and cause her to feel the effects more keenly. Moreover, woman, having forged a closer link with love, is endowed with a deeper instinctive perception of the divine aspects of creation than is man; and for this reason she may achieve greater evolutionary progress in a given incarnation—all of which would account for the necessity of beings having to experience female, as well as male, incarnations during their evolutionary cycles.

CHAPTER SIX

Children from the Stars

Gloria W. first reported this experience to me in 1969. At that time she was a sophomore honor student majoring in chemistry.

To focus, to relax, to escape from the pressures of the classroom, Gloria found great release in camping. Her father had been an avid camper and had taken her along on trips ever since she was a child of five or six.

Gloria was attending a college in Wisconsin, and she took great delight in exploring the beautiful forest and lake country on weekends. She found the out-of-doors the most perfect place she knew to recharge her batteries for the rigors of classroom discipline. Gloria was determined to excel in her chosen field of science, and she kept at the books relentlessly during the week. When the weekend came and while others partied, Gloria found her strength and her rejuvenation in the woods.

She remembers clearly that the incident occurred on a very lovely May evening when she lay in her sleeping bag beside an idyllic forest stream. She was letting the sound of the waters moving musically over the rocks lull her into a restful sleep when her half-opened eyes saw in the night sky a streak of light slash across the darkness and appear to come down in the forest. Her first response was that she was observing a meteor, but she knew that the light was too bright for any but the largest meteors—and that these were so rare as to not even be worth considering. She put all thoughts out of her mind, thinking that it had been merely a trick of the eye, perhaps due to her own fatigue.

She had drifted off into a light sleep when, about 3:00 A.M., she became aware of movement in the forest and what she at first identified as the sounds of men and women singing. She assumed that since the words were indistinguishable she was hearing a group of half-drunken college students thrashing through the woods.

Gloria looked up from her sleeping bag and saw lights bobbing in the forest. The musical sounds continued, but now in full wakefulness she was able to determine that there was not a chorus being sung, but independent solos being hummed. It was almost as if a moving opera were wending its way through the forest, she recalled, the cast members humming their lines.

Expecting at any moment to see the forms of half-stoned college students burst from the brush, she was astonished when she saw smallish figures with unusually large heads emerge. In the shadows created by their flashlights, they had grotesque, gargoyle-like appearances. They looked to Gloria as if they were snakes, or amphibians, two-legged toad-like beings, moving through the forest. Their eyes were unusually large, and they appeared to have no lips, just a straight line for a mouth, again reminding her of a reptile or an amphibian. She could see no discernible nose but only slits.

One of the individuals was carrying a boxlike object in his hands. It was emitting strange, crackling sounds, and rays of light were shooting out from it. Suddenly the entity carrying the box stopped and began to let out high-pitched squeaking noises, which Gloria took to be sounds of excitement. The other entities gathered around the one carrying the box, and they began to look around them into the darkness of the forest.

A shiver moved through Gloria when she received the clear impression that somehow the technology of that particular box had detected her presence, and now the toad people were looking for her. She lay there in utter silence, afraid to breathe, feeling sweat trickle down the back of her neck. A dry feeling entered her throat, and she was just at the point of leaping up from her sleeping bag and running into the woods. She decided that she would do her best to escape the strange entities that were emerging from the forest. She could count as many as six or seven of them, and her conclusion was that it was better to try to make a break for it and try to outrun them. After all, she had been a medalist in her high school track days. She should be able to outrun the short-legged, amphibious creatures.

Just as she was about to bound from her sleeping bag and head for the woods, a gentle touch on her shoulder caused her to turn, aghast with fear.

Her terror quickly subsided when she found herself looking into the face of the most beautiful man that she had ever seen. He had long golden hair that touched his shoulders. His eyes seemed large and loving, and even though in the darkness she could not determine the eye color, she somehow felt that if she could behold him in the sunlight, they would be a deep, beautiful blue.

All of her fearfulness subsided; all concern for the smallish, alien figures was removed from her thoughts. Although the stranger did not speak, he began to caress her, to put his arms around her.

Gloria felt herself responding in a very open way. Although she was not a virgin, her sexual activities had been very limited because of her disciplined approach to her studies. She had been dating a fellow chemistry major who had taken advantage of a scholarship in Europe. She had not seen him for seven or eight months, and she had

had no sexual contact with anyone else in the interim. As this stranger took her in his arms, she felt herself becoming sexually aroused with an intensity that she had never before experienced.

She remembers making love beside the forest stream. She had not a concern in the world. She had no apprehensions, no misgivings, and the shock of seeing the alien beings had been completely removed from her consciousness.

The next thing she knew she was blinking against the sun of high noon. Somehow she had slept through the morning. Her mouth felt dry, and she had a slight headache when she awakened.

She looked about her, hoping to see the beautiful, silent stranger of the night. She felt a flush move over her as she remembered the passion and the gentleness of their lovemaking.

Then the memory of the strange reptilian entities struck like a blow in the stomach. She jumped up immediately, prepared to run, if need be, prepared to fight for her life against the grotesque beings.

She could hear only the sounds of the forest birds, the sound of the stream. There was no sign as she looked around her camp that anyone or anything had been there the night before—especially creatures so alien and strange. Even if those beings had come from another planet, they had not left the slightest sign to indicate that they had visited Earth.

Three months later the campus health service physician told Gloria that she was pregnant.

Gloria insisted that she had not had sexual intercourse with anyone other than the mysterious stranger for several months prior to their act of love beside the forest stream. Against the strongest opposition of her parents and her closest friends, Gloria decided to carry the child full term and to keep it.

That child is now a brilliant seventeen-year-old college sophomore majoring, as did his mother, in chemistry. Gloria named him Ramar, for she "knew" that was his father's name. Ramar has unusually large brown eyes and a lovely golden complexion. His hair is a bright yellow-blond.

Gloria married two years after her graduation from college and has given birth to twin daughters. Ramar is loving and gentle with them, she reports, and the girls, now eleven, respond to him in a very beautiful way.

Ramar seems to be gifted with psychic abilities far beyond the ordinary, and he has excelled in all of his studies ever since he entered school. Ramar is also a superb athlete, although he specializes in track and field events of individual excellence rather than in team sports.

Gloria is convinced that the beautiful man with whom she made love was perhaps one of the reptilian entities that had the ability either to alter its own appearance or to create the illusion of being a human male. However she became pregnant, she is certain that Ramar's father was not a native of Earth. She is convinced that she has given birth to an alien child, a Star Child, whose appreciation of Earth exceeds that of most of those whose heritage is totally of this planet.

Ramar, in Gloria's opinion, is a citizen of the universe, filled with love, filled with compassion, and filled with a responsibility toward all living things.

Gloria admits that she used to long to see the father of Ramar once again, regardless of whether he would appear in reptilian form or in the idealized and masculine human form. She says that she no longer maintains such a desire. She has gone about her life, and she feels that she has been privileged to serve as one of those individuals who may literally be the mother of a new race composed of earthlings and Star Persons.

BORN OF UNCERTAIN PARENTAGE

Shelly B., of Minneapolis, Minnesota, said she was born on June 15, 1950, at 1:00 A.M. to a mother who had had three previous miscarriages and who had to remain flat on her back for the first three months of her pregnancy to enable her to keep what eventually would be Shelly.

Shelly was told by her mother that at her birth she did not cry like the other babies did, but just uttered a few little sounds. Shelly wasn't red and bloody like the other newborn children were. Her mother said that Shelly refused to take her breasts, so she had to be put on a formula immediately.

Shelly was told by her mother that during her toddler years she never spoke much, but her mother always seemed to know what she wanted. She was late learning to speak, because she and her mother had developed telepathic communication. To this day, she and her mother "know what one another mean to say and often will say the same thing simultaneously."

When Shelly was five years old, she experienced a very bright light in her eye that awed her so much that she called out to her mother in the middle of the night. Shelly asked where the light had come from, but her mother couldn't answer her. She just didn't know what a light in the eye might be. Shelly continued to see night lights of different colors. At the same time, she became preoccupied with pixies, fairies, and elves, and she began to draw pictures of them. Some of the drawings were deemed so professional by her parents that they were displayed in art showings and in galleries in the local area.

At the age of eleven, Shelly became infatuated with a being from a parallel world. She drew a portrait of him and began letting him speak with her voice. She told no one about her fantasy, not even her mother, in whom she confided nearly everything.

All through her teenage years, Shelly communicated with this being consciously, allowing him to speak with her voice, until she reached the age of twenty-four, when she met the man who is her husband. She is convinced that this other-dimensional being had been her companion even before she realized his existence at the age of eleven.

When the motion picture *Close Encounters of the Third Kind* was released in 1977, Shelly felt a very powerful chord being struck in her heart. She and her son went to science fiction films together, for both of them yearned for the stars. There was something about *Close Encounters* and the scene of the great mother ship and Devil's Tower that particularly struck her.

Soon after, Shelly was visited in her sleep by the same guide who had begun appearing to her and speaking through her when she was eleven years old. She was pulled out of her body and led to an asteroid where the guide told Shelly who he was and what he was to her. He was warm and reassuring and loving.

Shelly states that she has suffered ups and downs, but through courage and determination has managed enough compassion to remain with the man whom she has married, a man who she feels is not her true husband—although she loves him with an understanding type of love that she feels borders on unconditional love. She knows in her heart that the guide who appears to be a being of great wisdom is her true husband, her true cosmic mate, as she has come to call him.

In 1986, however, she was told by him to give up her attachment to the idea that he was a physical being. When she did that, she experienced a period of spiritual growth that had been unequaled in her entire lifetime. She was introduced to new techniques in meditation, New Age thought, books on spirit guides, and channeling.

In January of 1987, she found what she finally believes to be her professional calling. Shelly decided that she was put here to heal, not just physically or psychologically, but spiritually.

Peggy, of Dearborn, Michigan, gave me a complete report of her daughter Sara, who came from an unplanned pregnancy.

She has now begun to question Sara's true origin and who the child's true father may have been.

Sara was closely monitored before and after birth because of the accelerated growth of her head. The doctors feared water on the brain because of the rapidity of her brain growth.

Finally, at one, Sara was X-rayed. The doctor told Peggy that while Sara's head was adult sized, there was no need to worry. She only had an excessively rapid brain development and was an exceptional child with potential genius ability.

In 1982, when Sara was three years old, Peggy was taking her out for an evening walk when Sara suddenly pointed out what looked like an extremely bright light in the evening sky. Peggy immediately scanned the sky for signs of the moon to see if that was what the light could have been. The moon was in its usual evening place.

Peggy grew intensely curious and began staring at the bright light. She imagined all sorts of things, such as weather satellites or balloons, but she suspected that what she saw was none of those. She grew frightened and stepped up her pace, holding Sara's hand, until she was more or less dragging the child along.

Peggy continued to glance over her shoulder as the light seemed to follow them around a complete street block. She gauged its distance in the sky with the trees and the moon. She knew that it was moving and following them. It was at this point that she panicked, picked Sara up, and began to run.

"The bright light followed us all the way home, and I ran into the neighbor's house, calling upon her to come and look at the light. I was then no longer afraid, as I had become aware that it had followed Sara and me for a reason. As people began to gather around and stare at the light, it continued to stay directly over our heads."

To prove the fascination that the mysterious light appeared to have for her and Sara, Peggy began to walk down the block. The light moved with them. Peggy went out of the house later, around 10:00 P.M., and drove in her car on a northbound errand. The light followed until it finally disappeared into darkness.

Shortly after that, faces began to appear to Sara in the night, and there seemed to be an assortment of unusual occurrences.

Two years later, Peggy had remarried and was living in another state. Sara and she were on their way to the laundry room in the apartment complex in which they lived when Sara pointed to the sky and said, "Mama, look, there is our light again." Peggy looked up, dropped her laundry basket, and shielded her eyes.

"The light was so bright it was blinding. It was directly over our heads. I knew that it was there for us again, but I wanted someone else to see it, too. We ran back toward the house, and I asked my husband to come out and look, also. There were several other witnesses.

"I felt a vibrant communication with this light during the whole time that it hovered above us. While no physical conversation took place, I knew I was receiving some sort of thought wave or implantation.

"My life and Sara's have never been the same from that point on. I actually welcomed the light this time and received something from it, as I am certain Sara did."

Peggy became convinced that she was a child of the light and that she was on a special mission here on Earth to help others. Sara and she suffered through many attacks of negativity during that period of time.

"I fought battles in the spirit with the aid of a guardian angel," Peggy said. "I experienced out-of-body trips led by a spirit guide with a golden aura."

During that same period of time in 1984, Peggy reported, Sara also had contact with spiritual beings who gave her counsel and would take her aboard a spacecraft that she would crudely sketch for Peggy. Sara began to recite sequences of numbers and to draw sketches of what looked like spacecraft containing computer keyboards and people with strange names.

The two entities who were seemingly the strongest contacts for Sara were *Pencilava* and *Oba*. Sara described them both to Peggy in great detail. Pencilava was a very pretty lady with hair piled on top of her head and who usually wore a pink gown. Oba was a godlike figure with longish white hair. He was a very commanding authority figure who was strict in his ideas about Sara's future. He gave her spiritual teachings, and when she tried to block these out because they kept her awake at night, he would reprimand her and tell her that this was her destiny. He would let her sleep, but he would always return—for he was her true "Daddy."

"I honestly think that this entity may have been her true father," Peggy said. She went on to say that Sara began to speak more and more of a new place to which she was taken aboard the spacecraft at night. Sara told Peggy that one day Earth would be fixed up just like new.

"Sara began to engage me in long conversations about our mission here on Earth. That we were to show people the way to the Light. Sara became concerned that so many people would be destroyed with the Earth as it progressed along its path of destruction."

Peggy tried to tell her that those people would be given a choice whether or not to trust in the Light and to accept eternal life.

Sara could often take the pain out of the back of Peggy's neck. Peggy was on medication for arthritis. She went frequently to chiropractors, and she had sought all kinds of medical help. For several years she had lived with excruciating pain in her neck that would seem to intensify with humidity changes. Sara would be able to soothe it with her little hands until Peggy could feel the pain slowly fading.

"Sara has always been overly compassionate for the plight of others, and she cries easily when she feels that she has hurt someone. In general, she is extremely emotionally sensitive."

Lila L., from Memphis, Tennessee, is twenty-two years old and has been communicating telepathically with someone she calls "Father," an entity who says he is her true male parent. Lila was an unexpected child, and now "Father" has awakened her to the knowledge of who her real parents are and where her home planet is. She says she cannot wait to go home. The entity who identifies himself as her father also states that Joyce, Lila's two-year-old daughter, is his granddaughter.

Theresa of Manitoba is a twenty-seven-year-old accountant who reported that it was not until she had been out on her own, away from home for three or four years, that her mother told her the following story:

Theresa's mother had been decreed sterile, and in spite of repeated attempts to become pregnant, it was to no avail. The doctors had suggested adoption, and Theresa's parents were considering this very strongly.

Then, Teresa's mother was awakened one night by a strange buzzing sound that she described as something like a metallic bee. She looked up and saw a bright light about the size of a soccer ball moving across the bedroom. Before she could say anything, before she could shout, before she could express fear or alarm, she felt herself entering an altered state of consciousness.

Dimly she remembered the light hovering above her husband. At that point, the husband, although he was asleep, became animated as if he were a marionette being pulled into sudden life. And although her husband never opened his eyes, he performed the act of love with her, and it was roughly nine months after that strange act of cosmic coitus that Theresa was born.

Theresa's mother then told her that on the way home from the hospital she became aware of a bright light in the sky above the automobile. As she walked into the house with the baby Theresa in her arms, the light seemed to hover at treetop level. The light, according to the mother, was witnessed by several neighbors and by Theresa's father. After hovering at treetop level for ten or fifteen minutes, it seemingly disappeared—or according to the mixed testimonies of the eyewitnesses, moved into the night sky at an enormous rate of speed.

Upon hearing her mother's story, Theresa said that she felt a little uneasy about exactly who her true father might be. She admitted that ever since she was a small child she has had a fascination with outer space and UFOs. It was during such a discussion of various science fiction and scientific concepts and the possibility that extraterrestrial visitations could be occurring that Theresa's mother told her the strange account of her conception and birth.

THE UFO WITH THE PINK NURSERY

Mary B., of New York, is convinced that she has a special mission here on Earth, but she is very confused as to what it might be. She reports that she began having dreams about UFO people when she was only five years old. She would see a large ship hovering over her parents' home, and then on a beam of light, entities would come into her room and look at her.

They seemed to be examining her, as if they were doctors. They did not speak, and their mouths seemed to be fixed in a permanent kind of half-quizzical smile. She was not alarmed; she was fascinated.

As she grew older, the examinations seemed to continue. She remembers having these dramatic UFO dreams at least every three to four months. Shortly after she turned ten, she remembers the entities coming to her, taking her by the hand, and apparently lifting her out of her body in a kind of astral dream. She remembers being taken to a lovely pink room where everything was soft, gentle, and loving. She recalls very pleasant music playing. She could not identify the music as any familiar tune, but it relaxed her, made her feel very comfortable. She felt somehow as though she were taken to a nursery.

Her most dramatic occurrence took place when she was thirteen years old. She was visited in her room by the entities, who stood back in the corner while a more human-appearing man approached her. She, in spite of her youth and her inexperience, knew that she was engaging in sexual intercourse. The man caressed her, but did not speak.

Within two months, Mary claimed she was pregnant. She was frightened. She was only thirteen years old. She became very concerned. She could not work up the nerve to tell her parents. She considered telling her school counselor, but she could not bear the shame and humiliation.

She knew that she had not had any type of physical experience with any boy her own age or any older man or boy. Her only sexual experience, she swears, came from the man who entered her room on a beam of light, the man who was accompanied by the same entities who had been visiting her since she was five years old.

And then she reports that the strangest thing happened. She had another dream in which the entities came to her room and again seemed to examine her. This time she felt a bit of pain, and remembers lying as if she were paralyzed while they performed an operation on her.

"I wasn't pregnant anymore," she said. "It was really weird. A short time after that dream, my periods resumed; and I knew, I knew with all my being and my inner conviction, that I was not pregnant.

"Several months later, I had the last of my UFO dreams. I dreamt that I was taken aboard this craft. Once again I was in that beautiful pink room and this time I was looking at a baby—a beautiful baby boy. The entities smiled and indicated that I could pick up the baby. I did so, and I had the strongest feeling that I was holding my own child. I caressed and held him and said, 'I love you.'

"Everything then became hazy. The pink room seemed to get smaller and smaller, and I seemed to be covered with a pink mist. I awakened back in my room, and I have never had another UFO dream of that type."

URI GELLER AND THE SPACE KIDS

Andrija Puharich was well known as a surgeon, an inventor, and an author. In June of 1947, when he was a resident in internal medicine and after he had worked for thirty-six hours in surgery, he saw a UFO as he came outside into the sunlight. The day before, he had seen a newspaper headline reporting the sighting near Mt. Rainier by Kenneth Arnold.

Some years later, Dr. Puharich began to turn his interests to parapsychology. He built two Faraday cages and ran experiments to test the talent of the famous medium, Eileen Garrett, who convinced Puharich of the reality of telepathy.

At about the same time, he met Dr. Vinod, an Indian who was lecturing in the United States. In December of 1952, Puharich was present when Dr. Vinod spontaneously channeled a message from "The Nine" while in a trance at a mansion in New York.

Puharich learned that The Nine are found in every great culture. They are reported in the creation myths; then the legends subside into one favorite god. The Nine do not represent polytheism but rather aspects of a unit. Dr. Vinod had never done trance communication before, but thereafter he began to work for six months doing trance work and bringing forth information and instruction in advanced Yoga techniques.

In 1969 a group of parapsychologists were attending a three-day conference at Wainwright House in Rye, New

York. They were sitting around engaged in shoptalk, complaining that nothing exciting was happening in parapsychology. One of the doctor attendees said, "We need a repeatable experiment for mind power."

The following day a letter came to the physicist Itzhak Bentov, informing him about Uri Geller and his mind power demonstrations in Israel. As the group learned about this young man, who was ostensibly able to bend metal, move watch hands, and drive while blindfolded, and who was a master clairvoyant, they became very excited. Puharich was sent on a mission to Israel to check out Uri Geller.

In their various mind experiments, Uri was able to replicate Puharich's drawings from another room exactly to scale. He also moved watch hands and bent and cracked a gold ring.

The ring was flown to Dr. William Tiller at the Stanford Research Institute. Because gold usually bends flat, rather than cracking, Tiller was quite excited by the artifact.

Under microscopic analysis, it appeared that part of the fracture showed a heated grain surrounded by a supercooled structure. Tiller was shocked and wanted to publish these weird facts. He also thought that astronaut Ed Mitchell would be happy to support such revelatory findings through the Stanford Research Institute.

In 1971, while Puharich was observing Uri entertaining troops at a military camp, he noticed a star rising above the horizon just as the young Israeli began to talk. Then, as the evening's performance was completed, the star began to go back down.

Puharich noticed that the same thing occurred every time Uri was performing. He knew by then that the object was not a star. He was convinced that it was a UFO.

From time to time, Puharich and Geller went into the desert to entertain for the military, and almost each time, military personnel would pick up a strange signal on radar where no craft—either Israeli or foreign—was known to be.

When officers asked Uri what it was, Uri would reply simply that it was a UFO.

One day when they were driving to an airport at about 5:30 A.M., Puharich and Uri saw a giant UFO just above a hill to their left. It was blocks long, metallic, with a pod on top and underneath. The two military men who were driving the Jeep never saw it, even when asked if there was anything "over there." Puharich has recorded several incidents when he and Uri saw UFOs and others did not.

He began to experiment with Uri on the teleportation of objects, even animals, and on dematerialization. Puharich took apart his ballpoint pen, made some identifying scratch marks on the components, and put the pen in a wooden box. Uri passed his hand over it. Upon opening the box, it appeared that nothing had happened, but when Puharich picked up the pen, it felt lighter. Upon disassembling it, he found that the writing cartridge had disappeared. Uri did not know where it was, and after several days, the matter was forgotten.

One day, Uri called Puharich out of the house and, following some kind of internal direction, picked up his girlfriend and drove the three of them out into the desert. In a depressed plain at about dusk, they saw a small, twenty-foot-diameter UFO.

The trio started forward, but Uri said the others were to remain back while he entered it, which Puharich reports he did. After about twenty minutes, Uri came back out, sort of dazed, holding in his hand something that he said the beings were sending to Puharich. When the two of them examined it, they discovered the missing brass ballpoint pen cartridge with the markings that Andrija had scratched on it for identification purposes.

After such a demonstration, Puharich wanted to find out exactly *who* was working with Uri, and he asked permission to hypnotize the young Israeli psychic. Uri said it was impossible for him to be hypnotized, but he would go along with Puharich's wishes.

After a period of time, Puharich succeeded, and a voice began to speak through Uri. Once again Puharich was listening to a representative of The Nine. It had been eighteen years since Puharich had had such communication, and he listened intently as the entity speaking through Uri told him that the little craft that he had observed was run by robot extensions of higher beings who were in a larger spacecraft observing the actions. The large craft had been located in that particular region of Israel for thousands of years, and they had created a buffer zone in the Sinai so that no serious war would break out.

Puharich remembers clearly that it was in December of 1971, and the transmissions from The Nine changed both of their lives when a voice told them of an impending Egyptian attack on Israel and said that Uri must be the shield of that nation. Puharich and Geller were to go to the highest Israeli intelligence and inform them.

The intelligence officer listened to them and, instead of rejecting the information, gave them a list of questions, the answers to which were known to only a very few of the highest intelligence personnel. Puharich put Uri into a hypnotic trance, and the voice of The Nine gave the answers.

When the two of them returned to the intelligence officer, he said that he would get back to them later. In two days he returned their call and said that the answers had all been correct. Israeli intelligence wanted to work with them.

"Israeli intelligence is the best in the world," Puharich observed. "For example, they knew the precise time that Nasser would die, by analysis of his feces and urine, which they somehow were collecting. The invasion plans of Egypt were known, but Sadat, who succeeded Nasser, wanted peace by the end of December. It is thought that the extraterrestrials had influenced Sadat through Uri's and my meditation."

Puharich always kept a tape recorder with him as he was responsible for getting accurate facts together to bring back to the other parapsychologists and medical doctors. One day he noticed that the tape recorder began to record on its own. Upon playback, Puharich heard a voice that represented itself once again as The Nine.

In essence, what the voice said was that there was a spaceship, Spectra, over the Middle East. It had been there for thousands of years. Behind Spectra is Jehovah, properly Jehoova, the leader of Hoova.

According to Jehoova, there is a great distortion in the Bible. The entity insists that he said precisely, "I am not God; I am an extraterrestrial. My job is to seed life in this part of the universe, including Earth."

The tapes of Spectra gave some history of the evolution of this planet. At the time of Noah, the Hoovas had reseeded. About every 5,200 years they evaluate the quality of Earth life. If it does not measure up, they start over with the reseeding. The absolute rule under which they operate is that they must keep freedom of will available to the citizens of planet Earth.

There are beings who look like humans here on Earth whose mission is to counter the planet's awful negativity. There will be tough times ahead for humanity. There will be floods. There will be extreme cold. There will be little food. There will be cataclysms. All these things are in preparation for the transformation of Earth, and whatever negativity is being eliminated at this time is in preparation for greater cataclysms which will purge or cleanse the planet.

In his subsequent investigations with the "Super Kids," Puharich found that these young people are able to do dematerializations, bend metal, and perform the other type of phenomena that Uri Geller was so masterful in executing. Dr. Puharich worked extensively with seventy-nine of the Super Kids between 1974 and 1979. According to his findings, they came from twenty-four different planetary civilizations, and each seemed to have a special role in his own culture—and a particular reason why, at this time, they were to help humankind.

Many of the kids did not know *why* they were here on Earth, and some would forget even after Puharich had reminded them.

According to the information that Puharich has received, there have been three types of extraterrestrials in our present cycle of humankind. Beginning in 1882, there were the Preparers. Then appeared the Housekeepers. From about 1947 until the present time, the entities that have been interacting with humankind have the particular job of carrying out the spiritual transformation of Earth and introducing methods of neutralizing nuclear radiation. Not many of them will remain here by the year 1990.

The seeds for the new root race are being implanted now. They are being put into human mothers, and later the fetuses will be removed for actual birth elsewhere—a sort of virgin birth. The Earth mothers may think they have experienced a false pregnancy or that they were just getting a bit fat. The infants will be removed from the mothers after some months of pregnancy and will be brought up in extraterrestrial craft. Later they will be returned to Earth.

Puharich is convinced that we have three major jobs ahead of us on Earth at the present time:

1. We need to develop an energy that is absolutely clean.

2. We need to sponge up the cancer, the herpes, the AIDS.

3. We need to engineer our development into better human beings.

The ways to do this are to clean our physical system by not abusing drugs or alcohol . . . to kill the negative aspects of ego . . . to learn how to handle sex as sacred . . . and to find our opportunity to serve others and to stick with it, without being afraid of being poor, unbeautiful, or unpopular.

ALIEN RENDEZVOUS IN COLORADO

A forest ranger named Rick, who is stationed in a very remote and desolate area in the mountains of Colorado, reports that in the winter of 1982, he received a radio communication that flashing lights had been spotted along a little-used road up in the mountains and that this could be a large truck that had skidded in the snow. Since no radio message was received from any trucker, Rick feared the worst. In subzero February temperatures, a truck driver slumped unconscious in his cab would soon freeze to death.

Rick set off by snowmobile to the area where what appeared to have been flashing lights had been spotted by a helicopter the night before. A heavy snow had fallen three days prior to the alleged truck accident. Travel was hazardous for any but the most experienced drivers.

Rick spent over an hour crisscrossing the area where the helicopter had reported what could have been a truck's headlights. There were no tracks, no sign that any vehicle might have slid off the rudimentary country road and found itself somewhere in a large snowbank.

Rick radioed in the negative results of his search and suggested that if a truck had been stranded in the area, it had somehow managed to correct its own problems.

He was heading back to his post, a remote cabin that he maintained with another ranger, when he saw strange footprints leading into a stretch of forested area. The footprints appeared to be made by someone walking barefoot. Next to the trail of footprints was a smallish, almost pointed, shoe imprint.

Puzzled by such signs, Rick got off his snowmobile and began to walk toward the clump of trees to which the strange markings led. He was about to enter the thickly forested region when a smallish man stepped out with one hand raised, palm open, in the universal gesture of peace.

Rick was startled as he looked at an entity roughly five feet tall with a very large head, enormously large eyes, and a very blank, expressionless face. He was also astonished to see that in subzero temperatures, the smallish man wore only what appeared to be a tight-fitting, greenish-colored coverall.

If he feared for the comfort of the smallish man, Rick was completely taken aback when a blond woman with brilliant blue eyes, dressed in only a very thin, diaphanous gown, appeared, moving slowly, almost languorously, out of the forest. The light material of her garment seemed to swirl about her as if it were creating its own energy field.

Rick could not help but follow the entire line of the lovely lady's body down to her feet. Without any apparent discomfort, she stood barefoot in the snow.

Rick told them to get on the back of his snowmobile before they froze to death. It was Rick's thought at that time that he had run into a couple of weirdos from some traveling carnival. Perhaps there had been a stranded truck that he had overlooked. Perhaps a carnival rig had slid into deep snow, and the strange, bulbous-headed dwarf and this beautiful show girl were wandering dazed in the snow.

As he prepared his snowmobile for the trip to his cabin, he was interrupted by the sound of a strange, crackling noise, almost like static electricity. When he looked toward the strange duo, only the woman remained there. The man seemed to have disappeared.

When he asked the lady what had happened to her companion, she merely smiled. Rick wanted to go on a search for the dwarflike man, but the compelling eyes of the mysterious lady seemed to move inside of his head. He could think of nothing other than getting her safely to his cabin, and with an unprofessional nonchalance toward the safety of the other stranded person, he indicated that the lady should get on the back of his snowmobile. Together they sped off into a night that was rapidly becoming colder.

When Rick arrived at the cabin, he found that his partner was still out on an assignment. He made the woman a cup of coffee. She sipped at the cup slowly, looking at the contents as if she had never before seen coffee. Rick caught her putting her finger in to test the brew as if she could somehow pick up the liquid with her fingertips and drink it that way.

Rick remembers that she never finished her coffee. She smiled, very aggressively motioned toward his bed, and made it very clear that she wished to join him there.

Rick felt a moment of awkwardness, even apprehension. Maybe this beautiful loony would, as in the slasher movies, suddenly pull a knife on him. But the more he looked into those marvelous blue eyes, the more peaceful, the more tranquil, and the more sexually aroused he became. Rick admits that he put up little resistance and that he did make love to the strange woman three times that evening.

When he awakened at what he assumed was the next morning, the woman was gone. Rick sat up suddenly, fearfully.

Again he reprimanded himself for having behaved in a totally irresponsible manner. He had let a woman dressed in a very light gown wander off into the freezing temperatures with nothing on her feet and with nothing heavier than what appeared to be gauze to cover her body. He remembered making love to her, and he felt even more embarrassed that he had violated his position of responsibility. And what of her strange companion? Would he find him frozen somewhere in the snow?

At the same time, Rick began to wonder what had happened to his partner. Next to his friend's coffee cup he found a note scrawled that his partner had come back, found Rick in a deep sleep, touched his forehead, felt that he must be in a fever, and decided to let him sleep. He was on another assignment, and Rick was astonished by the date, for it was now two days later. Rick really began to feel ashamed of his actions. He had taken advantage of a lost and nearly helpless woman, and he had been asleep for two days.

As memory came back to him, the whole experience seemed more bizarre with every increasing recollection. First of all, he began to get a clear image that he had not taken advantage of the lost woman at all. If anything, she had been direct and very forceful about what she wanted from him. A truck was never found, and neither, to his great relief, were the frozen bodies of the woman or the little man.

Rick had nearly forgotten about the incident. Then, about eighteen months later, he was washing dishes in his cabin on a bright spring afternoon and he happened to glance up. To his astonishment, he saw the woman standing next to the large pine tree in the front of his house. She was smiling and dressed in the same swirling, diaphanous gown. In her arms she held a small, golden-haired child. Rick remembers that she pointed at him, then at the baby, then her stomach. She gave him another broad smile and walked back into the woods.

Rick said that he was unable to move for several minutes, but when he did, he dashed out the door and tried to find the mysterious woman. He could discover no trace of her, but it was clear to him that she was telling him that he had fathered the child that she carried in her arms.

This was a woman who had appeared impervious to the cold, who could walk barefoot through the snow in subzero temperatures, and who could apparently come and go as she chose—whether from some other world or some other dimension, Rick does not claim to know. He remains uneasy with the conviction that he has fathered an alien baby.

Every time he looks up at the night sky, Rick says, he wonders if his child is living on some other world. Or if the child might be growing up on some alien base, one day to return to claim its Earth heritage.

CHAPTER SEVEN

Contacting Their Own Kind

I have already recounted my own humanoid close encounter in previous works, so I will only summarize it here.

When I was a child, not quite five, I witnessed a most extraordinary occurrence. It was on an October night in 1940.

I remember clearly that I had been sitting on the edge of my bed, looking up at a harvest moon, when I heard the sound of someone walking outside on the crisp autumn leaves. Since my parents, my three-month-old sister, and I lived on an Iowa farm two miles east of town, the footsteps of anyone approaching the house—especially after dark—indicated an occasion of some importance.

When I heard a tin washtub being dragged from the pump at the well, curiosity and a small sense of alarm prompted me to get off my bed and walk closer to the window to investigate. I was astonished, rather than frightened, by the sight of a smallish man settling the tub beneath our kitchen window. The peculiar little fellow was dressed in a one-piece coverall, something like the kind that Dad wore when he worked on machinery; but the stranger's outfit was very tight, almost molded to his body.

The kitchen curtains were open, and the light from the kerosene lamp illuminated the little man's head and upper body as he raised himself on tiptoe to peer in at my mother and father. I could see his very large, round skull, two pointed ears, and long, slender fingers as they grasped the windowsill.

I didn't really see his eyes until he must have sensed that he, the watcher, was being watched, and he turned to look at me from a distance of no more than seven feet. Although we were physically separated by a windowpane, the transparent barrier did nothing to refract the tingle of shock and surprise that I received from those enormous, slanted eyes with their vertical, reptilian pupils.

I drew in a sharp gasp of air, and my heart began to thud. But the more I looked into the shadowed depths of those eyes, the calmer I became. They were already disproportionately large for such a small head, but now they were expanding even more. They seemed to grow larger and larger, more and more enchanting, until the next thing I knew it was morning.

A potent seed had been activated within my psyche. No amount of adult persuasion could alter the evidence of my personal experience. We were not alone in the universe. There were other intelligences walking about who resembled us, who seemed curious about us, and who might even care about us in some way.

A PORTRAIT OF MY ALIEN FRIEND

In July of 1987, my wife Sherry and I, together with our friends Patricia Rochelle and Jon Diegel, were attending the Outer Space Art Show of Luis Romero in Sedona, Arizona. I was both pleased and shocked when I beheld a painting of a particular alien. Incredibly, I now stood face-to-face with the nearest facsimile that I had ever seen of the entity of my childhood encounter.

I experienced a very strange kind of déjà vu in my solar plexus. "This is the closest I have seen to the entity that I witnessed as a child," I told Luis, who had heard the story of my encounter at a recent lecture.

Luis smiled and said, "I rather thought that you would have such a shock of recognition when you saw this work." He went on to tell me that he had done the painting to the specifications of a woman who had had an encounter with the alien as a child and who had continued to interact with that particular being.

He had no sooner told me that he had painted that work for a contactee when the lady under discussion walked into the art show. Her name was Sharon Reed, and to my astonishment, she had spent her childhood in West Bend, Iowa. Bode, Iowa, where we had our family farm, and West Bend are neighboring small towns, and while Sharon and I had never met each other while growing up, we obviously had a very fascinating friend in common.

As we compared notes, we discovered that we had both seen the entity at approximately the same time, and while we cannot say it was exactly the same night, it had to have been within a few days of each other's experience.

Contact with the entity had changed Sharon's life as the visitation had changed mine. She had practiced meditation for many years, and in the summer of 1975, while she was sitting very much at peace inside of herself, a vision appeared and seemed to block out all awareness of anything else in the room around her.

The room in which she was sitting was the living room of her California home, a beautiful room with many large windows for viewing the lovely scenery outside. This room was on the top floor of her trilevel house, and the view to the sky, the lake, and the hills was totally unobstructed by the other houses.

At first when the vision appeared before her, Sharon felt shock and surprise at what she was feeling and seeing. The vision appeared as a large light penetrating the room. Within the body of light was a large screen for viewing, much the same type of screen on which one would watch home movies. Appearing on the screen was a long, strangely shaped object that seemed to be floating in space.

She had never seen such an object before, and her mind was asking a thousand questions. She felt no panic or fear within her, only calmness and curiosity. She sensed that this object was from another world, another place in time, and the next thing she remembers is that she was standing *inside* the cigar-shaped object.

She was directed to a central glasslike dome standing alone in the center of this room. Inside the dome was a large book that seemed to be supporting itself in midair. She could not touch the book, because the glass dome was completely encasing it.

She was aware of beings who directed her attention to the book. They were polite, and they seemed eager to have her view it. As she stood before the dome, the book opened, but the letters of the book were unrecognizable to her.

Nonetheless, as she scrutinized the strange script, she could feel the knowledge of what they represented flowing into her mind. She could not interpret their meaning consciously, but she knew that somewhere inside of her the knowledge was being deposited.

As she studied each page, the book seemed to come alive, and the pages began turning over slowly. She continued to read each page as it was presented to her.

She stood transfixed in front of the glass dome, viewing the strange book and reading its messages. She became more aware of the beings who were in the room with her, and as she gazed at them, a strong feeling of calmness developed within her. She felt a strong bond with the entities.

When she finished reading the book, she said thank you to the beings who were so kind and intelligent. Although they appeared physically very different from her, that difference didn't seem to matter—for they had shared something very important together.

Sharon knew inwardly that what she had been given to read was implanted within her. Somewhere within her there was a space that had been reserved just for this experience and for the storage of what she had been given to understand.

VISITORS FROM GILANEA

Since that time Sharon has had many contact experiences with the beings. "There have been many stories published that tell of cruelties being done to Earth beings who are abducted and taken aboard a spaceship. Some people have told of painful tests that have been performed on them. This is an example of how fear can influence the mind and cause persons to perceive what has happened to them in an incorrect analysis."

The visitors from Gilanea (the place by which they have identified themselves to Sharon) have not brought physical harm to any Earth being. However, the space beings know that Earth occupants might bring harm to them, simply because fear, when manifested in an Earth being, can cause acts of violence and panic.

Sharon was told that when the entities with whom she had contact approached earthlings, they did so with compassion, patience, and wisdom. Earth beings were sent thought waves that penetrated their brain with a soothing, anesthetic-type effect. Fear was erased so that the Earth beings would not be harmed by their own fears, which could cause stress on their physical bodies. Earth beings were then able to meet the space visitors on an informative and mutually beneficial basis.

Many Earth beings today who have previously been contacted by the space visitors of Gilanea would have no fear of meeting the entities again. They have met positive space beings. Those who have encountered the benevolent entities know deep within their inner selves of the beauty, love, and compassion extended to them, and they know the space visitors to be highly special beings, whose love for all creation is a love that is unknown here on planet Earth.

Sharon Reed was told that the space visitors dwell on seven different dimensions. Their advanced technology would enable them to control completely Earth beings if they so desired. However, their absolute respect for the laws of the universe governs all their acts, and they do not seek superiority and power through control. Their Planet Earth Probe Mission has been an extremely long and difficult task, and their many messages to Earth have seemed at times to be almost desperate. As a parent to a child, they hope that Earth beings are listening to them, but only time will tell if the human race has responded.

The entities of Gilanea told Sharon that they have been active participants in the role of protectors of the universe many times before. They are spiritual beings who do not fail. However, they indicated that they needed the support and the attention of Earth beings, for without it, they could only be of superficial help on a very limited basis.

The planet Earth, the beings from Gilanea have come to know, also very definitely needs the attention of its own inhabitants. The entities are always seeking ways to improve their communications with human beings, but *Homo sapiens* must learn to become an active participant in the preservation of its own world.

The beings told Sharon that they are presently working with many of Earth's children through mind coding. These children, if planet Earth survives, will grow to become individuals of great accomplishment. They will be individuals with a strong inner belief in *what* they are doing and with a full knowledge of *why* they are performing their particular missions. There will be great changes among the many cultures that inhabit planet Earth.

When Sharon asked the beings what right they had to interfere with the children of Earth, their response was the following: "Our right comes into being because these special children are our children, also. They were not brought forth from the seeds of Earth people, but from the seeds of spiritual survival that comes from the higher energy level of love. You are fortunate that there are so many of these children now on planet Earth. Even though these children will be dissuaded by many of their Earth parents because their beliefs will be so different, they will risk all to have their purpose and message survive.

"You will come to know who these children are in a given time. Their souls are of a high spiritual level. They do not know fear. They will be quick to hold out their hand of knowledge and share it with others. These children will keep the road to survival well lit so that others may walk that path without stumbling. These are the children of your God, and they are being born into Earth's families everywhere. They have been coming into being since your year 1955. Since your year 1974, these children have begun to arrive in greater masses upon the Earth, because the need for these children is greater than ever before in the history of your planet."

The space beings told Sharon that Earth parents were carefully studied by them as potential parents for these seeded children of mercy. "The sincere ability of experiencing love in its purest form was the most crucial test each potential parent had to meet."

Sharon was told that she herself had been one of the selected parents, and it was explained to her how she had met the criteria set forth by the beings from Gilanea.

"Your search for a Superior Being that the churches call God seemed never to end, but then you finally found it—that knowledge of God within yourself. You dared to be different, because you had an inner knowledge that told you that you were correct. You have suffered the pangs of ugliness and despair that seem always to hover over the planet Earth. You have a strong desire to help humankind, yet you feel confused as to what steps to undertake to achieve that desire. What you do not know is that you are helping in a most profound and in a most important way by being a parent to God's children of mercy. Your own life is being blessed by these children, and you are truly in every sense their parent. Your aura shines brightly with love, patience of heart, sincerity of giving, and a closeness to your God. You are special, because you have been entrusted with God's most important creation—a child. Do not fail, for failure on your part will only bring failure to you, not your child, and you will experience the pain and torture of that child. Give your child true love, for that is the greatest nourishment you can give—and this nourishment shall return to you through your child."

"THEY WANTED TO INCARNATE THROUGH ME"

Shortly after Doriel, of Chicago, Illinois, saw a UFO, two beings appeared to her:

"One, a female, was named Leita or Leia. The other, a male, was named Gamal. They told me that they wanted to incarnate through me.

"I asked them, why me? They told me because I was one of them. They said that I could provide the right environment for them.

"I have a daughter now. I gave her Leita for a middle name. I was told six months before I was pregnant that a child would be coming to me. I practiced Tantric Yoga exercises prenatally and had natural childbirth.

"After she was born, I told her what solar system she was in, which galaxy, and that I would be her guardian.

Physiologically speaking, I had been told that I would be unable ever to become pregnant."

SELECTING "THEIR KIND"

Sometime in 1975, Karen of Grand Rapids, Michigan, dreamed that she heard a voice coming from the hill behind her house:

"I got up and put on my robe and followed the voice that was calling my name. I walked over the hill, and in the field behind it was a UFO, a very large one. I saw three figures standing beside it. I walked up to them, but I could say nothing. I felt they had control over me.

"They explained why they were there and why they wanted me. They wanted me to give birth to one of their kind. I had been selected because I was of their kind, also.

"One of the three men, who was standing on my right, came up to me and slowly started slipping off my robe. I tried to move, but I could not. The man on my left stepped forward and started touching me. All I could do was cry. They told me that they would not hurt me, so I should not worry.

"As they helped me with my robe, I could hear them speaking to me. Their mouths were not moving, so I knew that they were using telepathy. They told me that I could go back and that they would be contacting me at a later time. The next thing I knew, I was at my patio door. [In her communication, Karen added that her daughter, who was by then five years old, had been observed levitating. The child also spoke intimately of relatives, deceased before her birth, and she had already outlined her future life as a healer in a hospital.]

"These men were dark complexioned, with slightly slanted eyes. They were small of build and stood about five feet to five feet, four inches tall. They wore two-piece suits with belts around the waist. They had boots on their feet with their pants legs tucked inside.

"On their belt buckles they had some sort of symbol. It looked like some kind of bird in flight."

REMEMBERING CHILDHOOD CONTACT EXPERIENCES

When he was younger, Olof Jonsson, the psychic engineer who was chosen to participate in the Apollo 14 ESP experiments with astronaut Ed Mitchell, was hailed as the "Master of the Law of Gravity." He began rolling catsup bottles across the kitchen table by psychokinesis (PK) when he was just a boy. Jonsson had developed PK to such a degree that European parapsychologists insisted that the young man spend every available moment submitting to their laboratory tests.

As a child, Olof remembers that he received instructions from beings who "may have been the same entities that so often represent themselves to small children as fairies and wood sprites. . . . Their skin color varied from bluish green to golden brown to a shade of gray. It was they who began to tell me wonderful things about the universe and cosmic harmony. . . . I am still convinced that they are friendly and intend to help man as much as they can without interfering in his own development and free will."

Uri Geller is another who distorts the laws of ordinary physics when he bends metal with a gesture and materializes and dematerializes objects at will. According to Dr. Andrija Puharich, "Uri believes there are intelligences in the universe that have these powers, and that what he has is a tiny mirror image of theirs."

Uri Geller remembers playing outside one afternoon when he was three and experiencing a "flash" inside as well as outside of his head. The flash was very vivid and real to him, and he can recall that there were no rain clouds or lightning in the sky. "I didn't think much about it until much later, when things began to happen," Geller told the editors of *Psychic* magazine (May/June 1973). "I connect the flash with my ability."

Mary M. is employed by a television station in Santa Barbara, California. Things have begun to happen in Mary's life that have caused her to reassess who she really is and her true place of origin. She has begun to think seriously about where "home" is for her now. Although she is working as an executive for a television station, she has had a number of unsettling experiences in the last two years.

Recently, while she was reassessing some of her losses by going deep within her inner self, the planet Venus kept coming to her attention. In her dreams, she was told to buy and study quartz crystals, which she began to do. At the same time, she began recalling that when she was five years old, she had visitors in her room at night. She began to remember either being taken aboard a spacecraft or having arrived on Earth aboard a spacecraft at that time.

When Mary was twenty-six, her life was in great danger. At that time, a beautiful white figure appeared in a shaft of blue light. The entity told Mary telepathically that she should not worry about being harmed, because she was being protected. That experience strengthened her conviction that there is no death after leaving the physical body. And it also showed her that she had friends in the spirit world.

Since that time, Mary has had other visitations. She has had precognitive dreams, and she has experienced conscious astral projection. She has felt every earthquake in California (those that measured 4.5 and above on the Richter scale) twelve to twenty-four hours before it occurred.

Art, a printer from Phoenix, was three years and nine months of age when he encountered his "Lady" on the beach one afternoon in June.

"Much has been lost and forgotten due to adults talking down my stories and giving me excuses for what I 'thought I saw,'" he told me. "The only thing I remember vividly was the loving feeling and the thought: 'You have so much to learn and so little time.' She repeated that statement over and over again—never actually speaking, but rather transmitting the thought while smiling."

Art's parents couldn't keep him in the house after that encounter with the Lady. He would crawl out of his window and lie on the porch roof most of the night. When they moved to a farm near Barstow, Illinois, he would sleep in the trees so that he might be nearer the stars.

Then, one night when he was eleven, he and a friend were standing on a small bridge when a very bright object swooped down from the east and hovered over their heads.

"I tried to get my friend to look at it," Art remembers, "but he seemed in a trance, so he did not see it."

Art began a quest that evening that has taken him to nearly every part of the United States in search of more meaningful answers to life.

After his wife died in 1974, they continued an "exchange of wisdom" until his Lady arrived and took her to a "place of learning."

The Lady's last appearance to Art took place on April 17, 1979. Since then his teachings, warnings, and guidance have come from a male entity.

Art receives his transmissions "preceded by a pulsating buzz, a bright flash of light, then it's down to business— brief and to the point."

Art's son was born two and a half months premature, just as he himself had been. The boy walked and talked intelligently at the age of nine months. Then, when he was about four years old, he received nightly visits from the Lady for three weeks in June.

Because Art was receptive to his son's discussions of the visitations and because he encouraged the recitation of the transmissions, the child told him that the Lady had said that he "had much to learn in so little time." Supported by Art's interest, the child went on to relay the entity's teachings about death, the life beyond, and spiritual protection.

As a child struggling for survival in Germany during World War II, Christa had her life saved twice by three entities who guided her to safety.

Once she was trapped in the cellar of her uncle's home, the survivor of a direct bomb hit. The entities, who at that time appeared as three bright lights, contacted Christa's grandparents in a mutual dream and caused them to insist that digging be resumed.

Later, the entities manifested in human form and motioned her off a bridge just before Allied bombs obliterated it.

Christa saw her first UFO when she was thirteen and living in Great Falls, Montana. The experience began her longing for the stars.

"I knew as I saw this UFO that I did not belong here on Terra [Earth]," she wrote. "Longingly, I wanted to go to my true home beyond our solar system. That night I was told in my dream that the time was not yet, and that I must grow and have patience."

In February of 1961, Christa "lost" five days of her life: "I remember a bright light, a control console, and revolving crystals. . . . I had knowledge of a planet that was my true ancestral home."

She awakened fifteen miles from Great Falls. Five days had passed—although it seemed as if only five minutes had elapsed.

IS RH NEGATIVE ALIEN BLOOD?

Two charming sisters, Bonnie and Mabel Royce, approached me after one of my lectures on the West Coast and asked if I were an Rh negative blood type. When I responded that I was, they became very excited, declaring that I then could be a descendant of the ancient astronauts.

To Mabel and Bonnie, it is a mystery where Rh negative people come from. Most people familiar with blood factors, they said, admitted that people with this blood type must at least be mutations, if not descendants of a different ancestral line.

If Rh negative people are mutations, then what caused the mutation? And why does such a mutation continue with the exact characteristics? And who was the original ancestor who had the Rh negative blood?

It has been demonstrated, they went on, that the majority of humankind, at least 85 percent, has a blood factor in common with the rhesus monkey. This is called Rhesus positive blood, usually shortened to Rh positive. This factor is completely independent of the A, B, and O blood types. Mabel and Bonnie pointed out that in the study of genetics, we find that we can only inherit what our ancestors possessed, except in the case of mutations. We can have any of numerous combinations of traits inherited from all of our ancestors. Therefore, if man and ape evolved from a common ancestor, their blood would have evolved in the same manner.

Blood factors are transmitted with much more exactitude than any other characteristic. It would seem that modern man and the rhesus monkey may have had a common ancestor sometime in the ancient past. All other earthly primates also have this Rh factor, but, Mabel and Bonnie emphasize, this leaves out people who are Rh negative.

If all members of humankind evolved from the same ancestor, their blood would be compatible. Where did the

Rh negatives come from? If the Rh negatives are not the descendants of prehistoric humans, could they be descendants of UFO beings who came here eons ago?

Warming to their thesis, the sisters pointed out that all animals and other living creatures known to science can breed with any other of their species. Relative size and color makes no difference. Why, they ask, does hemolytic disease occur in humans if all people belong to the same species?

Hemolytic disease is the allergic reaction that occurs when an Rh negative mother is carrying an Rh positive child. Her blood builds up antibodies to destroy an alien substance in the same way that it would eliminate a virus. This process destroys any subsequent Rh positive infant she may conceive. Why, Mabel and Bonnie ask, would a mother's body reject her own offspring? Nowhere else in nature does this occur naturally.

The highest percentage of Rh negative blood occurs among the Basque people of Spain and France. About 30 percent of these have RR (Rh negative blood) and about 60 percent carry one (R) negative gene. Only 15 percent of the average population is Rh negative, while in some groups the percentage is even smaller. The Mizrahi Jews of Israel have a high percentage of Rh negative blood types, although most other Asian people are only about 1 percent Rh negative. The Samaritans and the black Cochin Jews also have a high percentage of Rh negative blood.

Mabel and Bonnie asked the pointed question: Could Basque people be the descendants of an extraterrestrial colony? Or to go even further, could the Basque region have been the site of the space beings' original colony on Earth?

The origin of the Basque is unknown. Their language is unlike any other European language. Some theorists maintain that Basque was the original tongue of the Book of Genesis. Some believe it was the original language spoken by humankind and possibly the speech of the "gods" who came in ancient times.

Quoting Genesis 6:2: "The sons of God saw the daughter of men that they were fair; and took them wives of all which they chose." Who were the children of these marriages? In Genesis 6:4, it states: "God came in unto the daughters of men, and they bare children to them, and the same became mighty men which were of old."

Mabel and Bonnie found it fascinating that the word *blood* is mentioned more often than any other word in the Bible except *God*. Those two words, they said, could be found on almost every page—*blood* and *God*. The blood of the gods, as they pointed out.

Mabel Royce said that they had searched in vain for scientific proof that Rh negative blood was a natural earthly occurrence. Instead, she said, she had found proof that Rh negative had not evolved on Earth in the natural course of events.

"For many years people have been searching for the missing link," Mabel said. "Could the true missing link actually be man himself? The unknown link between Earth and the Stars—hybrid man? Man may be the missing link between primate and extraterrestrial. It seems inconceivable to me that those working on the evolution theory have overlooked this possibility. How can they state that these people are lacking a factor contained in all other earthly primates including the naked ape, and not ask why? What other characteristics are common among these people that are uncommon to others? Is there a real difference other than just a different blood type? The Rh negative blood, which appears not to have originated on Earth, may prove to be a major factor in demonstrating that humankind is a hybrid."

Mabel admits that examining Rh negative blood may not be the entire answer, but she insists that it is a key toward unlocking the genetic puzzle of our true heritage.

"My research has shown that the majority of those men and women with psychic powers and abilities also have Rh

negative blood," Mabel said. "Most psychics and faith heal-
ers have this blood type. Strangely enough, many of those
doing research into the ancient astronaut theory and other
phenomena also have Rh negative blood.

"Why is there such a large percentage of Rh negative
in these unusual fields? Could these people have a vague
memory of their true origin?"

HE REMEMBERED A SPACE LANGUAGE

If the alien abductors should one day decide to declare
themselves openly to the world as agents of a superior culture, what would our governments do? Counterattack with
nuclear ammo? Or submit to the space beings and allow
them to take over? If their weapons are as superior to our
warheads and missiles as their heavier-than-air vehicles are
to our jets and rockets, a war between the worlds might
result in extinction of our species.

Dr. W. John Weilgart drew upon his universal consciousness to declare that humankind should neither attack
nor submit! "Our species must talk to them, write to them.
Certain of the UFOnauts might be appearing to warn us,
to advise us, to help us. We must try to communicate with
them."

But how could we do that? one immediately count-
ers. And how could our uninvited guests learn the more
than seven hundred languages of our illogical little planet?
Should they bother to learn the 600,000 words of even one
major Earth language? Should they spend months master-
ing the spelling and pronunciation of the English language
so that they may enunciate properly when they approach
the White House?

Dr. Weilgart was not one to set forth a difficult prem-
ise without offering a solution. The answer, according to
the Vienna-born psychotherapist, lay in aUI, the language
of space.

This language, aUI (pronounced ah-oo-EE), consists of thirty sounds, each a category and a symbol. The entire Logos can fit on a postage stamp and be learned within an hour. In Dr. Weilgart's words, "aUI is the language of space.

"It is the primeval language of the Logos of pure reason that takes our human race back before the confusion of tongues at the Tower of Babel. It is a transparent language in which each word becomes like a chemical formula of its contents—similar concepts sound similar, different ideas sound different, opposites are recognized. Word and meaning are one, and aUI can be radioed to the most distant galaxy. In this language of space there can be no trage-dies of prejudice. There are no synonyms or homonyms, no puns or double-talk."

Although he earned doctoral degrees in linguistics and psychology from the universities at Vienna and Heidelberg, in a single mystic moment Dr. Weilgart was given aUI, a new language for a new brotherhood of man, through a much higher awareness.

When he was a child of five in his native Austria, Dr. Weilgart had his first cosmic experience. A stranger in a star-strewn mantle appeared to him and said: "Thou must die, Wolfgang."

The child answered firmly: "I am ready." From this moment on he felt a new, cosmic life-stream had entered him, as if his former life had been dissolved, as if his former self had died. Wolfgang dictated to his older sister a drama, SaradUris (which, in the language of space, means that "good light" which comes "through the Spirit").

SaradUris is a spaceman savior who comes to Earth to guide humankind away from its wars and to bring peace through understanding. The rules of peace revealed in this drama are not only derived from sentimental love, but also from a justice based on the status quo as well as on a program of survival for those who are worthy and able to survive. "Creative Spirit, lives on still/Destroyers perish, while they kill."

Dr. Weilgart related his most crucial cosmic experience in poetic form as an introduction to his book aUI, *the Language of Space*.

The boy Johnny, seated at a mountain creek, is alerted to an alien presence by a strange, flutelike sound. There is a mysterious being, a butterfly with green, leafy wings and rootlike legs.

> It circled and sat down on my knee.
> The sounds it made were piped through its flute-
> 　　like body,
> and yet they wafted from far away.
> And then I knew that this thing must have
> come from somewhere else.
> And I began to understand that it was talking to me.
> Could we be friends?
> It showed strange signs,
> and put them together,
> and they made sense.
> And what I heard was from another place.
> I learnt
> The Language of Space.

Johnny learns that the being has come from a distant star, from where it had watched earthmen fight against the dragons of the sagas. "But now when nature's foes lie defeated at his feet, man turns his weapons against his brethren, and his rallying cry is still the slogan of hate, the word of conventional language."

> Your words give murder a beautiful name,
> and call a killer a hero.
> So if I gave you the gold of my wisdom
> wrapped in the burlap bags of your words,
> You would use it to club each other to death. . . .

As a young psychotherapist, Dr. Weilgart received a shock when he encountered a young boy in a mental hospital who addressed him in aUI. Among the words that the boy spoke was *bru* (b–together, r–good, u–man; i.e., a man with whom we feel good together—a friend).

Dr. Weilgart learned that the boy had been placed in the hospital because he insisted on speaking only in this language, which was unintelligible to his parents, teachers, and physicians. When the boy spoke in his mother tongue, he would tell of a visit to another world far from this one.

Dr. Weilgart went to the director of the hospital and told him that the boy was not sick, that his language made sense. The director ignored the young doctor and ordered electroshock treatment for the boy.

"The boy resisted the electroshock as much as he could," Dr. Weilgart recalled. "He told me that he feared that the treatments would make him forget what he had learned on the other star. In his struggle, he died, but with his dying words he said to me: 'You believe in me. You tell them about the language of space!'

"One cannot use aUI for puns, double-talk, slogans, or political invective," Dr. Weilgart pointed out. "Analysis into its 'chemical' elements dissolves implied insinuation into scientific proof of contradiction."

"How do I know the language of space?" Dr. Weilgart asked himself. "Is there a transmission of cosmic radiation into the ribonucleic acid of the genes or of the memory ganglia? Are some of us spacemen by heredity, radiation, or telepathy?

"I am a spaceman insofar as I feel one with the cosmos rather than with this strange planet, which earthlings have made a vale of tears. I have not what is typical of most earthlings, an ego, or power drive. Most earthlings would rather die or kill than let down their power ego. I feel at home when I look into the sky."

On January 26, 1982, W. John Weilgart's spirit returned home to the stars.

CHAPTER EIGHT

Alien Incarnations

Jeff M., of San Diego, has a distinct memory from the time when he was about five years old. He remembers standing outside, facing south, when he realized that he had come from "somewhere else."

"Associated with the place from which I had originally come were feelings of love, harmony, and longing," he said. "I longed to go back there, but I knew that I had a purpose for being here on Earth and that I would die before I could go back. I also knew, with some resignation, that I would live to attain a very great age before I would return to that beautiful place, my true home."

Jeff has known all of his life that he volunteered to come to Earth to help humanity. He has done this more or less in isolation of other helpers. Lately, though, he reports that he has been feeling a strong desire to connect with others like himself.

"I have intellectually known that they were out there all the time, but it did not seem right to search for them. I know that I have entered a new cycle of awareness for myself. I am now feeling a sense of renewed purpose. Both my spiritual growth and psychic abilities seem to be accelerating, and I am feeling a growing sense of urgency to find my true identity."

Jeff states that he has had multiple spiritual experiences and visions. He remembers one vivid experience in which he was driving with his girlfriend in a van near a large lake. Suddenly he sensed that there was a spaceship nearby, and it was coming to pick him up. Jeff became alarmed and thought to himself, "Not yet; don't take me yet. I haven't finished my mission."

The next thing Jeff knew, he and his girlfriend were on the ship, and beings dressed in flowing robes were ushering the two of them to a loungelike area where other humans were sitting around chatting.

"They took me to a part of the ship that had a console. The console had a viewing screen which at the moment was depicting a caravan crossing the desert in what appeared to be ancient Egypt. I got the impression that the viewing screen was used for monitoring and/or recording events that either have taken, are taking, or will take place on Earth.

"The most interesting thing was that the being who escorted me to this console said to me, 'I think you already know how to use this,' and sat me down as if I were expected to immediately go to work. Amazingly, I found myself manipulating various dials and levers and producing valid images."

Natalie, of Milpitas, California, was contacted first by a being of light who showed her the physical reality of their spacecraft. Then, on June 26, 1980, she once again encountered the same entities:

"They gave me a demonstration of their powers, which were awesome to say the least. They can make things appear and disappear at will. They also told me to start practicing telepathy, as I will need my *previous* abilities.

"I was informed that the craft that I had flown *in a past life* had crashed to Earth, because it had materialized at the wrong angle to counteract the planet's gravitational pull. I was being activated at this time in this life experience in order to help establish a direct communications link between their planet and Earth.

"They took me on a tour of the interior of their ship. . . . I felt that I should bow to them, but they told me not to. *They said that I had once been as they now appeared to me.* I got this overpowering feeling of love and goodness from them. Other beings on the ship sent me peace and love vibrations. . . . The beings glowed. . . . They are very advanced, but they still wish to assist us during our coming times of change."

Robert P., of Bridgeport, Connecticut, said that he was told that just after he was born, a four-foot-diameter gaseous ball came down in the family's backyard and rolled around, making a buzzing sound. Robert's mother thought it was ball lightning, but then the object rose up and soared out of sight, leaving no damage or marks of any kind on the grass.

When he was fourteen, the death of his mother caused Robert to go inward to a great extent. He remembers being different from other children.

Early in 1987, Robert contacted a well-known psychic in his area who advised him that he had been reincarnated many times, always in a religious lifestyle, always wishing to be of service to humankind. He said that Robert had a strong band of violet in his aura, which signified spiritual service. Robert was also told that one of his guides was an alien intelligence.

When he asked for more information about the entity, the psychic responded that Robert would recognize the phrase "Brothers beyond the Sun, without the color, without the race."

Robert reports that when he heard the phrase, it awakened something within him. He felt like crying and laughing at the same time, and he had chills all over. He stated that the movie *Close Encounters of the Third Kind* had the same emotional effect on him.

Arlene O., from Cincinnati, Ohio, has felt since childhood that she does not belong on this planet, and she has experienced dreams of another world, as well as what would seem to be past-life memories of having lived on another planet.

"I am adopted. I do know the name of my maternal mother, though I have no idea who my natural father is. I was adopted by the registered nurse who helped deliver me. The adoption procedure was done very quickly, and I have never been able to find out anything about my father. I have an unusual blood type—it is O negative—and I have been told that it is very rare.

"For about the past ten to fifteen years, I have had recurring dreams of being involved in a great experiment, which implies to me the idea of having been seeded here on this planet.

"In many dreams I have found myself on board a spaceship. In several of these I have been able to look back and see Earth spinning in space. I have also dreamt of being on a planet with a green sky and two moons.

"I am extremely sensitive to light, and I usually sleep with my head all the way under the covers. I can function on four to six hours of sleep a night and sometimes can go one or two days with no sleep.

"Sometimes I am troubled by the fact that I was such an unwanted child. My mother was only sixteen, and my father's identity has always been kept totally secret from me.

"I dream of the great experiment. It seems to me that the time is coming to awaken to my true mission, my great work, my actual purpose here on Earth."

Chris, from Alberta, Canada, was born the sixth of thirteen children to "good and honorable North American Indian people." She was a difficult home birth, and her brother followed only fifteen months later. She was given to her grandparents to be raised by them until she left home when she was seventeen.

As long as Chris can remember, she has always known that her Earth parents weren't really her parents. Although she loves her parents, her grandparents, her brothers and sisters, she has never really felt close to them. She has always been aware that they weren't her real family. Knowing she was different from everyone else always left her with a feeling of isolation.

When she was about four or five she was playing on a lakeshore near her home. Suddenly she looked up and felt a warmth and a beauty around her. There was a radiant white and gold light shining around her, filling her with overwhelming love. She got down on her knees and swore that her life would be devoted to God.

Now, as a grown woman, she has been "picking up UFOs psychically." Sometimes she is convinced that she sees the "Pleiadean type of craft."

She sees three UFOs around her home quite frequently. One is a large, saucer-shaped craft, and there are two smaller, cigar-shaped ones, which appear to travel together.

One evening in November of 1986, Chris was sitting quietly reading when she felt a sudden atmospheric change in her home. Standing magically in front of her was a small (about four and a half feet tall) clay-colored man in a dark green robe with a wide black belt worn diagonally across his chest. He didn't speak to her at that time, but she felt a tremendous love and calm emanating from him. She was overwhelmed by the feeling, and she knew that she had known the entity before and may even have been his mate.

All he said to her, in thought transference, was "Soon." This has happened many times since. Chris reports that the entity now comes with two other beings.

. One night she was preparing to sleep when she felt his presence, but it was different this time. She looked up and saw that two beings were accompanying the robed entity, standing beside and behind him. They were dressed similarly, but in beige.

Again he said to Chris, "Soon. I must not speak to you now. You are to fall asleep immediately, and you may come with us."

Chris was not afraid, and she remembers telling the entity that she loved him. In the morning she woke up feeling extremely excited. All she has ever felt toward the entities is a tremendous amount of love.

REMEMBERING PAST ALIEN INCARNATIONS THROUGH THE CRYSTAL SKULL

In volume 2, number 3 (1987) of *UFO Magazine*, the international forum of extraterrestrial theories and phenomena, produced by Vicki Cooper and Sherie Stark, Sandra Bowen revealed her own remarkable abductee experience.

Some people have indicated that crystal skulls located around the Earth may have come from Atlantis, Lemuria, or Pleiadean bases on Mars and the Moon. The crystal skulls are perfectly formed and polished representations of proper human dimensions and proportions.

Joshua Shapiro, who has worked with Bowen and Nick Nocerino in investigating the skulls, feels that there is a definite link between UFOs and the origin of the skulls. He is not certain whether UFO entities brought the skulls here and helped primitive man to manufacture them, or if the skulls are those of actual space beings who were crystallized. Shapiro indicates that almost everyone who touches the skulls feels vibrationally attuned to them and senses that there is some link between the skulls and outer space beings.

Sandra Bowen was standing in a naturally occurring energy vortex created by a bed of quartz crystals under a tree. She remembers suddenly seeing a huge spaceship with blinking lights appear above the tree. A ramp came down, and two beings began to glide with her up into the craft.

"When I first went up, it was very pleasant," Sandra told Sherie Stark. "It was like being told that my search to understand why I was here on Earth would finally end. I've always felt that there was something I should be doing, but I haven't really known what it is—and I have been praying to God for answers. The beings told me that I would have the answer to my question if I came with them and that I would be taken care of."

Sandra Bowen remembers being escorted into the spacecraft where there was a line of people along a rail, having their hands stamped with numbers. This was not the conventional kind of stamp, she recalls, one that you might get if you went to a dance or some public place; rather, it was a form of light that remained and became a part of one's being.

One of the attendants was about to stamp her hand when he looked at her and mentally communicated to her that she had to come with him. He took her to two other beings, who went with her to a room to meet with a large, kind-faced man draped in black.

"This being seemed like someone whom I could immediately trust and feel comfortable with," Sandra Bowen continued her story. "I almost had the feeling that he was representative of my grandfather. I was told that it had been agreed upon by me in the beginning (before I was incarnated) that the knowledge that was in the crystal skulls would come to Earth and that at the right time, I would be part of a plan. I said, 'Well, I don't even know what you're talking about.'

"He replied that I did know and that when the knowledge was activated, I would know what to do, that I am here now in order for them to remind me of my commitment and to help me bring about my mission.

"He said that when the time was right, the two space beings who had accompanied me to his presence would be with me every step of the way. They would tell me exactly what people to deal with, who to go to, what to say. All I would have to do was to relax."

The two beings, who seemed like her guardians, proceeded to work on her with different types of equipment, putting light in certain parts of her body, using a scanning machine, and so forth, to prepare her for her mission.

They took her to an amber cave and handed her a crystal whistle with blue and yellow feathers hanging from it. They told her that it would activate the crystal skull when she blew into it and that she would have such a whistle in the future. She was instructed that she should remember how to use the whistle and precisely what its function was. The space beings went on to tell Sandra that they would keep in touch with her through dreams.

In the next few months, Sandra stated that dream teachings revealed a past life when she worked with the thirteen original skulls in Tibet. She was shown and told how and why the skulls had been brought here.

"They told me that extraterrestrials had delivered the original crystal skulls to advanced societies. They had taught the Atlanteans the process of crystallizing people to preserve their experiences and knowledges when they died. The Mitchell-Hedges skull was actually the skull of an Atlantean priestess, who was so honored and loved by her people that they crystallized her to the third eye. They used the power of their minds to re-create the person's molecular pattern in an alchemical process."

Nick Nocerino, a paranormal researcher and founder of the Society for Crystal Skulls, told Sherie Stark that Sandra

Bowen's story seemed to be realistic. She exhibited all the physical and emotional components that one encounters when one talks to people who have been on UFOs or who think that they have been.

"Sandra experienced the lost time, the anxiety, the pain syndrome, the joy." Nocerino theorizes that Sandra Bowen's experience might have something to do with the opening or awakening of the brain that is not limited to our normal conception of time and space. Such experiences might occur in another dimension through a "hidden doorway."

A PRIOR LIFE ABOARD A STARSHIP

Vickie B., of Glendale, California, states that at the age of five or six, she had a dream of being on another planet. She still remembers how beautiful it was. The colors were vivid and clear, and there were more varieties of colors than there were on Earth. The main color of the planet, she recalls, was a glorious yellow.

"That may be difficult to imagine, but just as our main colors here—the blue of the sky, the green of plant life, the brown earth—blend to make this wonderful planet, so on that world did the colors blend in the most wonderful way."

When Vickie was ten years old, her father passed away. The loss of her father was devastating to her. She began to feel more and more alienated from people around her.

"I felt completely different. I could not think the way most people thought or talk the way they talked. I couldn't understand the little girls squealing when the little boys chased them, nor could I understand the chase. Later on, I couldn't understand changing boyfriends in a week or anything that was done only as a status symbol. I don't understand the value of gossip, and I've never understood group or racial prejudice.

"I've always felt so alone, with such a deep longing inside, as if I didn't belong with these people here on Earth, but I haven't known where I belonged. I still feel that way. I have people I love, sure, friends I care about, but I feel deep down that it may be many lifetimes before I can see those whom I truly know, those beings who are love. I don't know—maybe it's a memory of the beginning."

Although Vickie firmly states that she considers the Earth to be very beautiful, she has never thought of herself as an earthling or felt truly at home on this planet. During the last several years, although she has come to love the Earth even more, she has become even more convinced of her extraterrestrial origin. In fact, she thinks that all humans are.

She wonders if the Cro-Magnon may have been a leftover from some other world, starting over here on Earth. And could the Star People be the enlightened brothers from the same part of the universe? Could a deep sense of love have brought all people together again?

She has been told in a psychic reading that she spent three hundred years on Venus in preparation before coming to Earth and that she comes from a place in the stars called Sananda. Vickie has also been told that somewhere in space there is a very large spaceship in which there is a body that she occupied in a previous lifetime lying in a state of suspended animation.

In 1963, when Vickie was eighteen years old, she met Glenn, whom she later married. Glenn told her that when he was twelve years old he had been entrusted with some information from the stars. At last Vickie had found someone with whom she could communicate, someone to whom she could relate.

Vickie ponders frequently a dream that she had in August 1966 when she, Glenn, and their six-month-old son Eric lived in a small apartment in San Diego. Glenn was on a midshipman cruise, as he was in the navy at that time. Their apartment was a second-floor corner with a fire escape outside the window.

"One night I was awakened by three beings in the room with me. There were two men and one woman. They looked like us except that their skin had a fluorescent glow. The woman had long red hair and violet eyes. They did not talk, but they communicated with me telepathically. They said they wanted me to go with them.

"We left through the window where a starship hovered beyond the balcony. I went with them into the interior of the ship. We must have floated rather than walked. I don't remember my feet moving at all.

"They directed me to go into what I knew to be a mind probe chamber. There was a glasslike cylindrical tube, probably about three feet in diameter, that came down from the ceiling. Inside it was an instrument about eight inches in diameter. It looked something like a dentist's X-ray machine. There were no wires or anything visible.

"One of the beings stepped into the circle, and the tube came down to surround his body.

"Then the instrument touched the top of his head to probe his mind.

"I asked them why they wanted me to do this, and they said that there was some information that they needed, information that I had buried in my memory, and they could only get it from me. I had the feeling that it was some kind of formula.

"I was more than a little uneasy. If I refused to cooperate, they were going to force me. But from somewhere within me, I found a kind of power of my own. I told them they could not force me to enter the mind probe. I put up my own shield of protection. I used a kind of mental force to keep their minds from touching mine. I told them that they had to let me go, and I was able to move past them out of the ship.

"The next thing I remember is awakening in the dark in the corner of our tiny kitchen, crouched on my knees and scared half to death."

In yet another dream, Vickie experienced what she thinks now may have been a past life aboard an immense starship.

"I had a little cubicle, probably fifteen feet or so in diameter. There were instruments all around the inside. My job was to scan a section of the universe, looking for anything unusual. It was probably something like a radar unit. It seems to me that this ship was stationary, maybe like a space station.

"What happened in the dream is that someone came into my room and asked me if I knew how to turn on a power switch. I said that I did and left the room with him. There was a switch between an outer and inner wall which I activated, then I went back to my job."

Vickie was quite astonished when, in December 1976, Glenn began to channel a voice that identified itself as a being from another world. Vickie asked if she could use the name Alpha in referring to the entity, and the being said that that was fine since she needed a point of reference. Alpha explained to Vickie that he represented multidimensional entities from near the central Sun. When Vickie asked if they had bodies like humans, Alpha replied, no, they had bodies of light.

Alpha and other beings, which later channeled through Glenn, spoke of the Circle of Brotherhood, of being from a race that died from a genetic disease. They said repeatedly that humankind must unite their minds and join the Brotherhood.

Vickie said that one night she heard voices identifying themselves as entities from the planet Xerces in the star system Xandor. The beings declared that their home was three hundred million light years away. They told Vickie that they were coming to Earth. At the same time she was directed to go out to her backyard barefoot and stand with her feet apart and her hands raised to about ten o'clock and two o'clock so that her body became like a five-pointed star.

"In so doing, the energies of the Sun during the day and those of the stars at night are absorbed, and the body becomes a sort of receiver for these energies," Vickie commented. "The energies of the universe can be pulled into the Earth while the Earth's energies can be directed outward. As I remembered, this could be accomplished like a figure eight: through the left hand, crossing over the body, and entering the Earth through the right side, then out from the Earth and into the left leg, up through the body, to cross over and leave through the right hand. Another procedure is to bring in the energies of the sun or the moon or the universe through both hands and to send energy out the same way."

Vickie also received special instructions on how to shake hands. "It should be with the fingers pointed upward at a slant and palms together with one thumb gently wrapped around the other and fingers gently gripping the top of the hand."

CHAPTER NINE

Astral Abductions

Susan H., of Pennsylvania, witnessed a low overflight of a UFO when she was five years old. She stood on her front lawn, spellbound, as the craft came speeding over the rooftops in her neighborhood.

Today she has a degree in chemistry, and she is the technical manager of an industrial laboratory. She is married, with no children—a levelheaded woman with her feet solidly on the ground. And yet she reports a vision, a dream of being taken to a large group of people who are gathered in a strange city. The people seem concerned and are talking with great agitation about recent uncertain reports of odd weather patterns. Susan knows about the reports, but she is not anxious, because to her, they were normal and expected.

"I responded to the milling group in a very matter-of-fact manner, saying, 'Yes, things are changing—because *they* are here.'"

At this point in the dream she gestures upward to a clouded sky where two, possibly three, golden-submarine-type spacecraft have appeared. "I had been totally aware of their impending arrival."

Susan says that the vision shifts, and she is transported with a large number of people aboard one of the craft. "Each of us has a small collection of things with us in various types of knapsacks, suitcases, and so forth. We are in this huge room that seems to have some curves to it. We receive a message to go to the view port. All of us are unafraid, accepting, and expectant.

"We leave our stuff along the corridor and proceed to the view port. We are hovering high enough, so that a sense of the curvature of the Earth can be seen, but we are not so high that we cannot see that we are over the water with the shoreline in full view. We are told that they will begin a rotation of the spaceship and that we will see that there is nothing to be concerned about, such as falling down because of gravity. The ship begins to rotate. As we watch the shoreline move during the rotation, we are amazed and excited and feeling somewhat unnatural. We are astonished when we stay in one spot with respect to the motion of the ship."

"It was an incredible feeling, and one which I distinctly recall to this day," Susan said. "It was almost like the kind of amusement park ride that strives to disorient through optical illusion. After the rotation, we were told to make ourselves comfortable, and we returned to the corridor outside the view port and sat on the floor. I had the sense that we were going on a long voyage, but one which I knew was not going to take long."

Mary B. is a freelance writer from Salem, Ohio. At around the age of ten, she experienced an astral projection or an actual physical encounter. It is all hazy to her now, but she remembers at night seeing a large spaceship hovering about twenty feet off the ground in her backyard.

A door opened in the bottom of the craft, and a light beamed down. A woman in a white gown appeared to Mary. Mary, reared as a Roman Catholic, somehow assumed the woman was Joan of Arc. She was not afraid of the glowing woman, and the entity gave her a message that she still, as an adult person, cannot understand very well.

But Mary remembers that the woman said that she knew that she was lonely and that she would give her a playmate who would always be with her. The woman said that she would have to go back.

Mary cried because she didn't want the glowing woman to go. The woman stood in the beam of white light again and cloned herself—made an exact duplicate of her physical form. This duplicate was the woman, but as she would have been as a little girl.

When she was eleven, a man in a spaceship came and took Mary inside of it. She remembers that it was all white, except for a purple object on a low white table. Mary does not remember what the entity told her, but he had black hair cut in an ancient Egyptian style. The experience seemed to have been done by astral projection.

Mary says that "they" have often guided her and her husband to various places. Once they saw seven craft, including a mother ship. Two other witnesses confirmed their sightings, and the experience was reported to a local UFO investigating group.

In 1986, Mary and her children saw a large ship hovering over their house. It was shaped like a diamond with one end cut off, and it had hundreds of gadgets in its underside. It was slow-moving and appeared two hundred or more feet long. It disappeared as if it vaporized.

Mary describes the entities with which she has contact as "extremely good-looking humans." They have a light around them and they glow, but they appear to be solid.

Although she has been in contact with them for years, their age always appears to be roughly twenty-six to thirty years old. The hair of both male and female is cut in a page-boy style. The man has black hair and the woman's is a golden blond. The woman wears a long white gown with a round neck. The man wears a white pantsuit garment. Both of them are very kind.

"They are not permitted to tell me everything, but they give hints so I can get the answers to my questions. They help me with my life. I don't receive contact with them very often, but they are watching me. They give me advice, which I try always to follow."

Mary reported that once a UFO guided her husband and herself to a nursing home in the area. What a perfect place, she thought, for aliens to be in charge. It would be ideal. It would not be detected, and they would be able to gain a great deal of Earth information from the elderly people.

CLOSE ENCOUNTERS OF THE SPIRITUAL KIND

Dreams, astral projection, past lives. It is clear that many people are confronted by UFO intelligences in this way rather than visually or physically. But is this a valid tool for exploring the UFO phenomenon?

If it is possible that the human essence can soar free of the accepted limitations of time and space imposed by the physical body and truly engage in "astral flight," "soul travel," or the more academic "out-of-body experience," then it may well be that the paraphysical aspect of humankind may more easily interact with that paraphysical species we commonly identify as the UFOnaut. Indeed, many accounts that tell of a subject having been taken aboard a UFO may be descriptions of a mental/spiritual/nonmaterial experience, rather than an actual physical/material one.

Let us consider a parallel between UFOs and the legends that grew up around the old religion—witchcraft—in the mid-1400s. For centuries the Christian Church officially ignored the practitioners of the ancient religion, but during the very dawn of the Age of Enlightenment, when men were seriously considering the structure of the universe, certain members of the Church hierarchy suddenly became obsessed with devils and women flying through the air on broomsticks.

In his *Antichrist and the Millennium*, E. R. Chamberlin makes an excellent point that may be analogous to the aspect of the UFO enigma under discussion in this chapter:

> *Paradoxically, it was the Christian Church which, seeking with all its powers to combat the practice of satanism, gave that same practice a form. It was necessary to define witchcraft in order to combat it, and by so defining, the Church gave shape to what had been little more than folklore. Most of the elements that eventually went to make up witchcraft had long been content to dismiss them as mere fantasy. The legend of the woman who flew by night came in for particular scorn. "Who is such a fool that he believes that to happen in the body which is done only in spirit?" Such sturdy common sense was forced to give ground at last to a rising tide of fanaticism.*

While certain readers may consider out-of-body experience (OBE) even more tenuous an object of pursuit than UFOs, a number of research laboratories are employing the most sophisticated scientific devices in a serious effort to establish OBE as a very real aspect of what it is to be human.

In a report on out-of-body research at the American Society for Psychical Research (*ASPR Newsletter*, number 22 [Summer 1974]), Dr. Karlis Osis, Director of Research, wrote:

> *For the past two years, the ASPR Research Department has been fully engaged in exploring the question: Does the human personality survive after bodily death? . . . We have been following up our central hypothesis: That a human being has an "ecsomatic" aspect, capable of operating independently of and away from his physical body—an aspect which might leave the body at death and continue to exist. Can one, we asked, really leave one's body temporarily (as in out-of-body experience or OBE) or permanently (as at death)?*

After a detailed review of current experimental projects, Dr. Osis summarized the ASPR work by stating:

The OBE research proved to be a difficult task, mainly because the full phenomenon is rarely produced at will. Our results are thus far consistent with the OBE hypothesis. After fully exploiting the research possibilities described above, we may indeed hope to have evidence for the ecsomatic existence of human personality.

Thousands of men and women have been provided with their own personal evidence and proof of the validity of the "ecsomatic existence of human personality." There is an enormous body of literature dealing with OBE, and numerous accounts of the phenomenon are to be found in mystical and religious traditions.

Spontaneous OBEs most often seem to fall within eight general categories: (1) projections while the subject sleeps; (2) projections while the subject is undergoing surgery, childbirth, tooth extraction, etc.; (3) projection at the time of accident, during which the subject receives a terrible physical jolt and seems to have his spirit literally thrown from his body; (4) projection during intense physical pain; (5) projection during illness; (6) projection during pseudo-death, wherein the subject "dies" for several moments and is subsequently revived (Elizabeth Kübler-Ross and John Moody are authors who deal with this phenomenon in great depth); (7) projection at the moment of death, when the deceased subject appears to a living percipient with whom they have an emotional link; (8) conscious out-of-body projections in which the subject deliberately seeks to project their spirit from their body.

Now, it would appear, another category must be added: projection during which the subject feels that they have been taken aboard a spaceship and have interacted with an alien intelligence.

Consider this account of the kind of UFO-OBE that results in a recognition between strangers:

> When I answered the door, I saw a friend and a stranger standing there. The newcomer had a look of shock on his face. For most of the evening he kept staring at me; I finally insisted that I know why. He said he had had a weird dream about someone he'd never met before, and he had recognized me as the man in the dream.
>
> He said that in his dream he had been in a clearing with a lot of other people. They seemed to be waiting for someone or something. He did not know anyone there except me. He said that I smiled at him and made him feel calm and peaceful. He trusted me. Then everyone began to look up.
>
> The sky was clear and star-studded except for a large circular patch directly overhead. Then he noticed that there was a large oval object blotting out the sky. As he realized this, an opening appeared in the center of the object, and a blue-white light spilled out.
>
> He felt strange and he looked around to see how the others were reacting. Then he noticed that everyone was floating up toward the opening, one by one.
>
> He blacked out and came to in a dome-shaped room. The other people appeared to be awakening at the same time. Everyone had been placed in one of the chairs that lined the walls in three tiers. Across from them were electronic panels with flashing lights, dials, switches. In the center of the cabinets were two seats in front of what appeared to be control panels. Behind this area was a brilliant light. In the exact center of the room was a column, or pole, running through the floor and ceiling. A low railing, about three feet high, encircled the column.
>
> He looked at the other men and women, and they appeared to be as confused as he was. He felt as though someone were missing. Then everyone turned and looked toward the center of the room. There stood

a man in a close-fitting, one-piece silvery spacesuit
that covered his hands and feet. He wore a globe over
his head that obscured his features. "Welcome aboard,
friends," he said, as he reached up and removed the
globe. And the stranger said that it was me!

I have now experienced this sort of thing again and
again over a period of a year and a half. The shocked
stranger, the stare, the same dream, down to the most
minute details. After the fifth or sixth time, I began
thinking, "Oh, no, not again!" I can't say how many
times this has happened. I have lost track.

In discussing this dream-recognition phenomenon further, the following additional comments and details were
produced:

One person having a dream about a future meeting
with a stranger is an occurrence not at all unfamiliar in
the literature of psychic phenomena. But we are speaking
of a situation in which approximately a dozen men and
women experienced the same dream, identical to the small-
est detail, all climaxing with the meeting of the same man.
This bends the laws of chance out of all proportion. My correspondent wrote:

I had them draw a floor plan of the dome-shaped
room. Allowing for differences in artistic ability, they
all drew the same, identical floor plan. I then had them
mark the position that they had occupied on the three
tiers of seats, hoping that some of them would mark
the same position. But none did. Each one had a different location in which he or she said they had been
sitting.

I then asked them to describe the suit that I had been
wearing. Again, the descriptions were identical. Just
about every detail that I could possibly think of, I
asked, and they all agreed.

Here is another interesting facet: I asked each one of
them when they had had the dream, and none could
remember. This puzzled some of them. Surely they
could recall vaguely if it had been a week, two weeks, a

month. But they had no idea whether it had been the night before or a month before. Apparently, the dreams were not normal dreams.

Upon further reflection, this correspondent decided to relate an extraordinary "dream" that he had experienced during the summer of 1959:

> I was sitting in my lounge chair, reading, dozing. Yet I was tense, nervous, restless. Something seemed to be in the back of my mind that I seemed to have forgotten but shouldn't have.

> Suddenly I looked toward the door. I knew someone was on the other side. I laid my book down, got up, went to the door, opened it a crack, and looked out.

> There were two men there dressed in black. They could have been identical twins, they looked so much alike. They were dark complexioned with Asian eyes, but they were definitely not Asians. Remember, this was in fifty-nine, long before people started talking about "Men in Black."

> They never said a word, but I heard inside my mind: "Are you ready?"

> I don't know why, but for some reason or other, I was ready to go. Since it was very hot that night, I had stripped down to my birthday suit, so I reached for a pair of walking shorts. Again, I heard inside my head: "That will not be necessary. No one will see you." Strangely enough, that seemed to satisfy me.

> We stepped out into the hall, then instantly we were on top of a flat hill in back of the apartments. I was rather surprised that the scene had changed so quickly. I noticed the headlights of a car coming down the street, and I ducked behind the two men. I didn't want to be seen running around in my birthday suit.

> I heard some laughter in my mind: "We told you no one would see you. Try it!"

> I boldly stepped around in front of them, spread my feet apart, propped my hands on my hips, daring anyone to see me. The car, with a man and a woman in

it, passed a few feet from us. They didn't even look in my direction. It was as if we weren't there. That surprised me.

I turned to say something to my companions, and they were looking up. (This is the part of my experience that is similar to the "dreams" the strangers told me.) I followed their gaze and realized that something was hanging there suspended above us. As I did so, an opening appeared in its center, and blue-white light came tumbling out of it.

I felt a queasy sensation in the pit of my stomach, like when you are in an elevator or an airplane that is dropping too fast. I could see the apartment houses and the ground receding below us. We were floating up toward whatever that thing was. I blacked out as we were approaching the opening.

When I came to, I was lying on my side facing a wall. I rolled over on my back and sat up. I was in an oddly shaped room. The best way I can describe it is like a wedge of pie with the point bitten off.

The whole room was bare except for some kind of projection on which I was sitting. Everything seemed to be made out of a blue-gray material. While the walls were very hard, the surface on which I was sitting was soft, even though everything seemed to be made of the same material. The room was bathed in a soft glow, and there were no shadows anywhere, but there was no light source that I could see.

I heard a female voice say, "He's awake now."

I looked around to see if I could spot a speaker, a TV camera, or something, but again I saw nothing but blank walls and ceiling.

About this time on the short wall—the one bitten off the end of the pie wedge—a door appeared and opened. I could see into a hallway. Although the hall was dark, there was blue-white illumination that appeared as though it were coming from some great distance.

Two shadows flitted across the doorway. I couldn't tell anything about their shapes. The movement was too rapid and too distorted.

*But I got a mental impression, if you will, of two
people approaching—a man in the front and a woman
in the back—carrying a trayful of some kind of surgical
instruments and hypodermic syringes.*

*The next thing I knew. I was back in my apartment,
in my chair, reading my book. I gave a shudder and
thought how sleepy I was. I went to bed, laughing
about what a vivid imagination I had.*

This correspondent said that when he awakened the
next morning, he regarded the whole episode as a strange
dream. But when he reached for the book he had left on
his desk, he found that it had disappeared. For two days he
searched the apartment without finding the book he had
been reading when the bizarre experience interrupted him.

When his roommate, a special agent of the FBI, returned
from a trip, he challenged him to prove his effectiveness by
finding the missing book. The two men turned the apartment upside down, searching for the vanished volume.

"We started at one end of the apartment," he writes,
"and we moved, dusted, waxed, and cleared everything all
the way to the other end. We found things we had forgotten
about, things that we'd thought we had lost someplace, but
no sign of the book."

Then, about one week later, the two men suddenly
found the missing book on the edge of the desk, right where
the correspondent had left it:

*If this book did indeed vanish from our apartment and
reappear, the question is how? which goes back to that
dream sequence again. If it really happened, how did
I get back in the apartment, since when I pulled the
door closed behind us, it locked automatically—and I
had no key in my birthday suit!*

*If we assume for a moment that the dream sequence
really occurred, it brings up some interesting points.
For one, the possibility of teleportation. When I was
let out into the hall, we were instantly on top of the
hill. There was no time lapse. It was an instantaneous
thing.*

If we did teleport, why didn't we teleport directly to the UFO? The only explanation I can reason out is that somehow or other the UFO was shielded or had some kind of radiation that prevented teleportation and we had to be levitated inside.

When I dream, I usually know, even in my dream, that I am dreaming. But in this case, I didn't have that knowledge, or even that feeling, of a dream. It was dreamlike because I had so little control over my actions, but there was no sense of time lapse between the moment I was reading my book and the moment I looked up at the door. If I had nodded off to sleep in those seconds, I would assume that there would have been a feeling of change; yet there didn't seem to be any. The only change seemed to be that I suddenly lost control of myself and became more of a robot than anything else.

I think a lot of us have been controlled telepathically, somehow or other, by the UFO intelligence. That's just a feeling I have. I have no proof.

That's the trouble with the whole UFO phenomenon. What concrete evidence is there? All the weird things that have occurred around UFOs seem to be leading us more into the area of parapsychology.

CHAPTER TEN

Kidnapped by the Fairy Folk

It might be said that there is a close similarity between ancient accounts of humans allegedly being kidnapped by fairies and those abductees who claim that their adversaries were members of UFO crews. UFOnauts and fairy folk even resemble one another according to descriptions given by witnesses throughout the centuries.

There are numerous accounts of humans allegedly being taken by fairies into a strange, enchanted servitude and being kept for a period of time—or perhaps never permitted to return to Earth at all. There are accounts in which babies or grown women are stolen, but the cases that seem most similar to the UFO abduction cases are those in which there is a supernatural lapse of time in fairyland, when a man or a woman has not merely been prevented from returning to home and human society but has also been rendered impervious to the ravages of time and the awareness of its flight.

In *The Science of Fairy Tales*, by Edwin Sidney Hartland, published in London in 1891, the account is given of a shepherd who went out one day to look for his cattle and sheep on the mountain. After about three weeks, the search parties had abandoned hope of ever finding him again. His wife had given him up for dead, and it was at that time that he returned.

When his astonished wife asked him where he had been for the past three weeks, the man angrily said that he had only been gone for three hours.

When he was asked to describe exactly where he had been, he said that he was surrounded at a distance by little men who closed nearer and nearer to him until they formed a very small circle. They sang and danced and so affected him that he quite lost himself.

Near Bridgend, Wales, is a place where a woman is said to have lived for ten years with the fairy folk and who, upon her return, insisted that she had not been out of the house for more than ten minutes.

The Germans, the Irish, the Scots, the English, and the Scandinavians have no end to such accounts of wee folks interacting with people and stealing time. We find variants of these tales in Wales, in the Slavic countries, and in Japan and China. Stories are told of men and women who returned years—sometimes even generations—after they had stepped into a fairy circle and been enchanted by the sounds of the singing and dancing of the wee people. Additional anecdotes are told of those who took to themselves husbands and wives from the fairy folk and produced a hybrid of human and fairy individuals.

In Scotland the story is repeated of a man who went with his friend to enter his first child's birth in the record books and to buy a keg of whiskey for the christening. As they sat down to rest, they heard the sound of piping and dancing. The father of the newborn child became curious, and spotting some wee folk beginning to dance, he decided to join them.

His friend fled the spot, but when the new father did not return for several months, he was accused of murdering him. Somehow he was able to persuade the court that he should be allowed a year and a day to vindicate himself.

Each night at dusk, he went to the spot where his friend had disappeared to call out his friend's name and to pray. One day just before the term ran out, he saw his friend dancing merrily with the fairies. The accused man succeeded in grabbing him by the sleeve and pulling him out.

The bewitched man snapped angrily because his friend would not let him finish the dance. The unfortunate friend, who would face the gallows if he could not bring the enchanted man home, said that the celebrating father had been dancing for twelve months and that he should have had enough. The man, when rescued from the fairies' circle, would not believe the lapse of time until he found his wife sitting by the door with their year-old son in her arms.

Several tribes of Native Americans have similar stories of interactions with entities they call the Star People. According to legend, the Native Americans took Star wives and husbands, and the Star Beings took Earth wives and husbands. Native Americans also refer to the medicine or magic circle. If a man stepped in one, he could disappear for months or years or a lifetime.

The magic circle and the fairy ring, the twirling lights and the ethereal music, seem another version of the hovering UFO that leaves a scorched mark in the farmer's field. The fairies, according to the belief of most people, are a race of beings who are counterparts of humankind in their person, but "unsubstantial and unreal, ordinarily invisible, noiseless in their motion." They possess magical powers, but are mortal in existence—though leading longer lives than humankind.

Nevertheless, they are strongly dependent on humans, and they seek to reinforce their own race by kidnapping human beings. The fairies are of a nature between spirits and humans, but they can intermarry and bear children. Along with the lure of music and dancing and lights, there have been other ways to seduce human beings into experiments of crossbreeding and physical examinations, such as promises of wealth and eternal youth.

It would seem that these entities have universally been calling themselves by very much the same name. The Native Americans divided their supernatural visitors and companions into two categories: those glowing lights in the sky were the Star People; and those who inhabited field and forest were the Puckwudjinies.

Here we have one of those cross-cultural references that prove to be so thought provoking. *Puckwudjinies* is an Algonquin name that signifies "little vanishing people." *Puck* is a generic of the Algonquin dialect, and its exact similitude to the Puck of the British fairy traditions is remarkable. Puck, or Robin Goodfellow, is the very personification of the woodland elf. He is Shakespeare's merry wanderer in *A Midsummer Night's Dream*—"sweet Puck," who declares what fools we mortals be.

Puck is no doubt derived from the old Gothic *Puke*, a generic name for minor spirits in all the Teutonic and Scandinavian dialects. *Puck* is cognate with the German *Spuk*, a goblin, and the Dutch *Spook*, a ghost. Then there is the Irish *Pooka* and the Cornish *Pixie*. To break Puckwudjinie even further and concentrate on its suffix, we find *Jini*, the Arab *Jinni* or *Genie*, the magical entity of the remarkable lamp.

There are numerous Native American legends that suggest an interaction between the native people and Star Dwellers. Nearly every tribe has its accounts of "Sky Ropes"—ropes of feathers—that permitted People from Above to come to the Earth Mother and, on occasion, enabled men and women to fly to the clouds. Along with the magical ropes are tales of flying canoes, airships, and moons that descended to Earth.

Many Native American tribes believed that the stars were the homes of higher beings who had a connection with, and a mysterious relationship to, humans. Others held that the stars themselves were actual ministering intelligences.

Numerous tribes had accounts of warriors who found themselves enamored of Star wives and of tribes of women who had been enticed by Star husbands. Often the Native Americans found "magic circles" that the Star People had burned into the grass, just as their European brothers across the ocean were finding "fairy circles" that the dancing elves had tromped into the meadows during nocturnal revels.

The Chippewa have a legend that tells of a great "star with wings" that hovered over the treetops. Some of the wise men thought of it as a precursor of good; others, understandably, feared the star and saw it as the forerunner of terrible times.

The star had hovered near the village for nearly one month when a Star Maiden approached a young warrior and told him that she was from the winged star. They had returned from a faraway place to this, the land of their forefathers, and they loved the happy race they saw living in the village. The star, she said, wished to live among them.

The warrior told the council of this visitation, and representatives went to welcome the Star People with sweet-scented herbs in their peace pipes. The winged star stayed with them for only a brief time, however, before it left to live in the southern sky. As a token of its eternal love, according to the Chippewa, the Star People left the white water lily on the surface of the lakes.

CHAPTER ELEVEN

Genesis II

On Dec. 17, 1868, the *Los Angeles News* provided some nineteenth-century food for thought.

> Captain Lacy of Hammondsville, Ohio, had some men engaged in making an entry into his coal bank, when a huge mass of coal fell down, disclosing a large, smooth slate wall, upon the surface of which were plainly carved several lines of hieroglyphics. No one has yet been able to tell in what language the words are written. The letters are raised; the first line contains twenty-five. It is probable that they were cut in the coal while in its vegetable state and during its formation into coal.

The men discovered the wall with its undecipherable hieroglyphics about one hundred feet below the surface. If the letters were cut into the coal in its "vegetable state," as the anonymous reporter suggests, then we are back in the Carboniferous Period, approximately *250 million* years ago.

FOOTPRINTS IN THE STRATA OF TIME

In the early 1930s, Dr. Wilbur Greeley Burroughs, head of the Geology Department of Berea College, Kentucky, was guided to a site in the Kentucky hills where he was able to locate ten complete manlike tracks and parts of several more in Carboniferous sandstone. All the accumulated evidence indicates that they were impressed upon a sandy beach in the Pennsylvanian Period of the Paleozoic Era—which dates the humanoid impressions somewhere around 250 million years ago. Dr. Burroughs kept his work secret for seven years. One can imagine that he wanted every opportunity to study the amazing tracks.

"Three pairs of tracks show both left and right footprints," Dr. Burroughs told Kent Previette of the *Louisville Courier-Journal* many years later (May 24, 1953). "Of these, two pairs show the left foot advanced relative to the right. The position of the feet is the same as that of a human being. The distance from heel to heel is eighteen inches. One pair shows the feet about parallel to each other, the distance between the feet being the same as that of a normal human being."

Regardless of the tests to which Dr. Burroughs subjected the tracks, the results were always the same: the footprints were genuinely those of a bipedal creature. It positioned its feet like a human, had a heel and five toes, and walked exclusively on its hind legs.

At the suggestion of Dr. Frank Thone, biology editor of *Science Service*, with the concurrence of Charles Gilmore of the Smithsonian Institution, Dr. Burroughs named the originator of the mysterious tracks *Phenanthropus mirabilis* ("looks human, remarkable").

The Pennsylvanian Period was the age of giant amphibians. Could the tracks have been made by one of them? Dr. Burroughs thought it unlikely. "There is no indication of front feet, though the rock is large enough to have shown front feet if they had been used in walking." Dr. Burroughs was emphatic that the creatures, whatever they might have been, walked on their hind legs. Nowhere on the site were there signs of belly or tail marks.

Is it possible that ancient Native American artisans or more contemporary sculptors could have carved those footsteps?

A sculptor informed Dr. Burroughs that any carving done in that kind of sandstone would be certain to leave telltale artificial markings. Neither enlarged photomicrographs nor enlarged infrared photographs revealed any "indications of carving or cutting of any sort."

On May 25, 1969, the *Tulsa Sunday World* carried an article describing fossilized footprints found by Troy Johnson, a North American-Rockwell liaison engineer. Just a few miles beyond Tulsa's eastern city limits, Johnson removed earth, roots, and stone from an outcropping of sandstone to find animal prints—many of which he could not identify— and some distinctive five-toed humanlike footprints.

More than just a weekend dabbler in archaeology, Johnson had thirteen years of experience in study and fieldwork for, among others, the University of Oklahoma and the University of Arkansas. At the time of his startling discovery he had also presented papers on his finds to archaeological associations.

C. H. McKennon of the *Tulsa Sunday World* presented Troy Johnson's quiet arguments in favor of the footprints' authenticity:

> The chunk of sandstone containing the big prints is a massive weight of an estimated fifteen tons, which rules out the possibility of someone transporting it to the top of the hill. Also, the stone is of the same strata as other specimens of sandstone dotting the hilltop, indicating there was a monumental 'uplift' of the Earth's crust ages ago.

PREHISTORIC GENETIC ENGINEERING?

In 1898, H. Flagler Cowden and his brother, Charles, unearthed the fossil remains of a giant female, over seven feet tall, who they speculated was a member of a race of large primitives who had vanished from the face of the earth some 100,000 years ago.

Astonishingly, the Cowdens had found their giant woman in Death Valley, an area that, while desolate in modern times, may have been an inlet for the Pacific Ocean in prehistoric times. In the same stratum with the female skeleton were the remains of extinct camels, elephants, palm trees, towering ferns, and fish life.

One thing is certain: if the giant female was seven feet, six inches tall, then assuming the same kind of height ratio that exists in modern times, the male of the vanished valley paradise would have been eight feet tall.

It is difficult not to feel tiny prickles of recognition of genetic engineering when we learn that the Cowdens discovered a number of anomalous physical appendages and attributes not found in contemporary man. They noted several extra "buttons" at the base of the woman's spine, "and every indication betraying the woman and her people as endowed with taillike appendages." The brothers also found that the woman had canine teeth twice the size and length of modern man.

[The hasty will scoff at the suggestion that giant humans with tails once walked the Americas, but the *New England Journal of Medicine* of May 20, 1982, described the birth of a baby with a two-inch-long tail. The slender, tapered growth was surgically removed at Children's Hospital Medical Center in Boston. Dr. Fred D. Ledley saw the appendage as a "vivid example of man's place in evolution."

While noting that few tail cases have been documented in the latter part of this century, Dr. Ledley stated that the "well-formed caudal appendage" represented a "striking clinical confrontation with the reality of evolution."

Humans may have diverted from their most closely related tail-bearing primates twenty-five million years ago,

Dr. Ledley agreed, but "human genes still contain information for tail formation."]

There are many other examples of the remains of strange, humanoid giants being unearthed by construction or mining crews. Put together, these strange discoveries follow a pattern and lead to unsettling questions that have a direct bearing on the subject of UFO abductions.

Is it too heretical to suggest that humankind may have received some external assistance in its evolutionary trek?

Does it remove humankind too much from the center of the universe to wonder if there might not have been a world of experimental and developmental humans before our own?

Rather than asking the question of whether or not the UFO abductors are conducting genetic experiments with unwilling men and women today, the more powerful query is whether humankind may have been structured by UFO intelligences in the beginning.

Dr. Paul MacLean, one of the world's most eminent brain researchers, theorizes that during the course of evolution, humans have acquired three very different brains: a primal mind from reptiles, an emotional mind from early mammals, and a rational mind from more recent mammals. The very center of our brain is composed of the primitive reptilian brain, largely responsible for our self-preservation.

Dr. Carl Sagan made extensive use of Dr. MacLean's theory in his bestselling book *The Dragons of Eden*, and the triune brain hypothesis has stimulated numerous scholars in the social sciences. Is it really too much to suggest that there may be an even more dramatic, more *direct*, reason for our reptilian heritage?

In my more than twenty years of investigation into the UFO mystery, I have heard those human participants of close encounters with alleged UFOnauts most often describe the aliens as smallish, reptilian-appearing humanoids with disproportionately large heads. Once, as I listened to these eyewitnesses recounting their experiences, my memory provided me with silent confirmation, for my own visitor on that long-ago October night had a preternaturally large head for its size.

On September 8, 1981, biologist Dr. Thomas E. Wagner and his coworkers at Ohio University announced the first successful transfer of a gene from one animal to another—from rabbits to mice and then to the mice's offspring.

If we now have begun the process of engineering the transfer of certain traits from one creature to another, we will soon be able to create genetically new animals or to transfer a particularly desired trait within the same species. If we are on the verge of such an accomplishment, where does that put the UFO intelligences—who clearly are far ahead of us technologically?

In 1953, Francis Crick and James Watson launched a "biological revolution" with their discovery of DNA, the master molecule that contains the genetic code. In 1962, for their successful solution of the DNA puzzle, Crick and Watson shared the Nobel Prize for Physiology and Medicine with physicist Maurice Wilkins.

Today, Crick has boldly suggested that the seeds of life on Earth may have been sent here in a rocket launched from some faraway planet by "creatures like ourselves." Science writer David Rorvik explored such a provocative theory, together with other subjects, in an interview that appeared in the March 1982 issue of *Omni* magazine:

> At [a meeting in Soviet Armenia in 1971] we discussed the idea that uniformity of the genetic code makes it look as if life went through a rather narrow bottleneck. In addition, I came to realize that sufficient time had elapsed for life to have evolved twice—that is, a civilization capable of sending out rockets could already have come into existence at the time the solar system and the Earth got going. Leslie Orgel [a biochemist at the Salk Institute] and I, in collaboration, came up with the idea of a directed panspermia.

When asked by Rorvik to comment on precisely which things made the theory of panspermia attractive to him, Dr. Crick replied: "The easiest way to see that it's attractive is to realize that we might find ourselves doing the same thing a thousand or two thousand years from now, seeding life in the same way."

From the perspective of the UFO intelligences, the process of evolutionary trial and error may not yet be completed. They may still be monitoring the development of our species through the programmed process of UFO abductions.

CHAPTER TWELVE

Armageddon:
The Clash between UFOs
of Light and Darkness

Numerous UFO investigators have worried about the possibility that our world could be conquered by the infiltration of alien agents. It becomes very frightening to many people to conceptualize that humankind could be overcome and even destroyed by programmed men and women from within the ranks of their own species.

If there is indeed a kind of final confrontation approaching, an army of men and women could be ready to take actions that they themselves do not even understand—yet they would have no option other than to obey when the prearranged signal was given. Their reaction would be as those individuals who are under deep hypnosis. They would obey simply because they had been conditioned to obey at a prearranged signal.

UFOlogist John Keel said, "We have no way of knowing how many human beings throughout the world may have been processed in this manner, as they would have absolutely no memory of undergoing the experience; and so we have no way of determining who among us has strange and sinister 'programs' lying dormant in the dark corners of his mind.

"Suppose the plan is to process millions of people and then at some future date trigger all of these minds at one time. Would we suddenly have a world of saints or would we have a world of armed maniacs shooting at one another from bell towers?"

In his book *UFOs and Their Mission Impossible*, Dr. Clifford Wilson states,

> *If Armageddon to which the Bible points is indeed a final battle in which human and nonhuman forces alike wage that dreadful conflict to the death, this sort of programming is a real possibility and appears to be proceeding at breakneck speed across the whole of the world. It is reported that the term Armageddon has been used in messages to contactees and other end-of-the-world messages have been given. Is there a desperate preparation for a last-ditch stand by the forces of evil? A final attempt to thwart the plans of the Holy God against whom they have rebelled?*

Dr. Wilson is one of a growing number of biblical scholars who theorized that the Lord Jesus's warning that there would be great signs in the heavens at the end of the age could be fulfilled in the UFO phenomenon of this generation. Dr. Wilson does not truly fear that humankind could ever be conquered by negative UFO entities, however. It is his contention that the great master plan of God would not permit an invasion by hostile beings from outer space. Neither, Dr. Wilson feels, is there any evidence that outer space beings are necessarily a race superior to humankind.

> *If it is true that they are using men, women, boys, and girls to serve their own purposes, the usefulness of those who thereby become their servants is limited to one generation. They are unable even to double the life span, let alone give man the fantastic life span that*

*would be necessary to move through the planetary
systems of the galaxies that are beyond our present
knowledge. It would appear that these people need a
superrace, unfettered by the fear of death, to accomplish their purposes. To find such a race on this Earth
is a "mission impossible."*

HAVE AGENTS OF THE SUPERPOWERS SOLD EARTH?

In 1977 David Ambrose and Christopher Miles prepared an
original script for British television called *Alternative 3*. This
documentary declared that the superpower governments
had devised a plan to preserve a tiny nucleus of human survivors and that many people of high intelligence and exper-
tise in science and technology were mysteriously vanishing
off the face of the Earth and being taken to bases on the
moon where they served as literal slaves.

According to the documentary, it had been determined
by the superpowers that Earth would soon be unable to
support life, that our climate's recent strange behavior was
only a warm-up for the tremendous cataclysms to come. The
superpowers have been working secretly together in space
for decades, and their accomplishments in achieving bases
and in conquering the farthest reaches of planetary travel
have advanced far beyond that which has been released to
the public. Ultrasecret joint United States and USSR con-
ferences are held each month in a submarine beneath the
Arctic ice cap. Government agencies have been kidnapping
ordinary people and turning them into mindless slaves by
advanced brainwashing methods. NASA's reports of strange
things that the astronauts saw on the moon were suppressed by the superpowers in order to keep the masses
ignorant of the overall sinister plan.

Although the television documentary and the book version by Leslie Watkins, published in 1978, were both decried

as science fiction with no basis in fact, our own research has produced similarly frightening accusations that not only portray the superpowers as working together in secrecy on an overall master plan, but, circa 1947, joining in a deal with intelligences from outer space. It would appear from our investigation that there may be more than one "government" existing in this nation and that there is an overall conspiracy of world leaders to maintain a policy of UFO secrecy. Certain militarists and industrialists form the actual power structure of the United States as well as the other superpowers.

Many sincere and conscientious public officials may not have any idea of the vast range of secret workings that are being conducted between this second government and the UFO intelligences. According to evidence produced by informants who have worked for various government agencies over the past forty years, there now appears to be an explanation—albeit even more terrifying and sinister than the others presented in this book—as to why the UFO abductions may be taking place. Certain UFO entities have apparently been given carte blanche to conduct various types of experiments on the human population. This is done in an effort to determine what type of hybridization can be accomplished and also exactly what type of stresses the human body and psyche can survive.

Witnesses exist who claim to have seen the bodies of the aliens who died in the Roswell, New Mexico, UFO crash landing in 1947. Although this particular event has been pooh-poohed in UFO and scientific literature ever since, the persistent research of such investigators as William Moore and Stanton Friedman has revealed that the event quite likely did take place. In our own research, my wife Sherry and I have spoken to scientific personnel who claim to have been on military duty and to have examined the bodies.

If these informants can be believed, it would appear that certain government officials at that time made a deal with the UFO intelligences to gain scientific and technological knowledge that would cause a great leap in our accomplishments as a species. Regretfully, these exchanges did not greatly benefit the average person, except in certain spin-off and peripheral side projects and byproducts. The principal thrust of this deal was made to profit only a few greedy people high up in the government.

This is not to suggest, according to our informants, that the entire government or any particular administration has sold us out. The second government, perhaps the more powerful one, has, for its own benefits, bartered with the aliens. What has occurred now, according to our informants, is that these greedy, power-hungry individuals have outsmarted themselves. Thinking that they could trust the aliens with whom they were dealing and thinking that they were their peers, they have now discovered that the aliens have been using them for their own ends, and that a mass invasion of Earth is imminent—and would have already taken place but for the intervention of other, more benevolent beings.

It seems that some kind of "Interplanetary Council" has decreed that a planet's biological and technological evolution should not be interfered with in any way. The very fact that certain alien entities made a deal with certain Earth government officials to exchange scientific knowledge for minerals, water, raw products—and even for human beings to be utilized in extraterrestrial experiments—totally violated the council's dictum that Earth's natural evolution should not be accelerated. The benevolent beings, so say our informants, are doing what they can to halt the invasion of Earth and to right the vast number of wrongs that have already been dealt to the human species.

Certain of our informants have described US military bases, now abandoned, that were being used to create "interdimensional tunnels" whereby aliens could enter the Earth's atmosphere with greater ease. Unspeakable types of experiments with young people were said to have been conducted. Large numbers of street kids and, in some cases, young homosexual males were apprehended or lured to areas where certain experiments in teleportation or mutations were conducted.

One informant told us that one of the tunnels below an abandoned military base contained the skeletal remains of several hundred young males. In his explanation, these were the street people, the runaways, the stray kids, that simply disappeared and were never really missed, except in those few instances when there were concerned relatives.

Bright youngsters, in a bizarre kind of a brain drain, were lured into government projects in ESP, mind control, and mind manipulation. It was the mystery of such an astonishing brain drain that led to the *Alternative 3* documentary on British TV. Those researchers, while trying to discover what was happening to our best and our brightest young people, stumbled upon the revelation that the superpowers were conspiring to create a new world of their own choosing and their own manipulation. What has since been discovered is that the superpowers made a deal with the alien intelligences who declared themselves to our leaders after the Roswell incident in 1947.

Even as this book is being prepared in September of 1987, documents have surfaced that would indicate, at least through circumstantial evidence, that presidents from Truman and Eisenhower to Reagan have been made aware of the crashed UFO in Roswell and the alien bodies that were found!

A secret document, known as MJ–12, was released in June 1987 by researchers William Moore, Stanton Friedman, and Jaime Shandera. Originally issued on November 18, 1952, MJ–12 refers to alleged events during the 1947 to 1952 period, highlighting an ultracovert body called "Majestic 12." The document seemingly was prepared to brief Eisenhower on the state of in-depth government UFO investigation at that time, and it begins by describing the 1947 Kenneth Arnold sighting of UFOs near Mt. Rainier in Washington State. The shocker is the reference to the military retrieval of a crashed UFO in Roswell, New Mexico.

Attachments to MJ–12 include such artifacts as a letter dated September 24, 1947, ostensibly written by then President Harry S. Truman to Secretary of Defense James Forrestal. According to the information in the document, the material contained therein went to several high-ranking government officials.

William Moore stated that the investigators have subjected the document to rigorous scrutiny in order to rule out the possibility that it might be a hoax. "MJ–12 has passed almost every test," Moore said. "The whereabouts of the officials named in the document have been satisfactorily verified. An independent expert examined the type style, saying that it came from an Underwood typewriter that would have been in use at the time. If this document turns out to be misinformation, then I submit it is one of the best and most organized misinformation schemes ever."

SOWERS OF CONFUSION AND DISCORD

The *Ahrimanes*, according to Persian and Chaldean tradition, are the fallen angels who were expelled from Heaven for their sins. They desire to settle down in various parts of the Earth, but they are always rejected. Out of revenge, they find their pleasure in tormenting the inhabitants of the planet. According to Persian legends, the Ahrimanes finally elected to inhabit the space between the Earth and the fixed stars, and there they claim their domain, which is called Ahriman-Abad.

Military and aviation historian Trevor James Constable has come to the conclusion that it is the Ahrimanic powers that are seeking to wrest control of planet Earth from its human inhabitants. Constable believes that Inner Space and not Outer Space is the invasion route that has been chosen by the Ahrimanic powers.

"A fifth column inside the human mind makes external force unnecessary," Constable says. "Washington has already been invaded from Inner Space. Moscow was long ago occupied by Inner Space invaders. Continuing ignorance of these malefic, invisible, but all-too-real forces will bring disaster not only to America but to human evolution and to the destiny of man as a free being."

Constable reminds us that the Ahrimanic powers have, for centuries, held as their goal the total enslavement of humankind. If they are unopposed, the Ahrimanic entities will overwhelm humankind and make evolution wholly a matter of their control.

"Man will win or lose the battle for Earth itself, for he is at once the goal of the battle and the battleground. The choice man makes—the extent to which he utilizes balancing forces to neutralize Ahrimanic assault—will bring him victory or defeat. The stakes in this battle are not the territory, commercial advantages, or political leverage of ordinary wars, but the mind and destiny of man."

Free will dictates that humankind has the right to decide whether or not life on Earth will be surrendered to the Ahrimanes, and Constable underscores that the ominous threat that the challenge presented by the Ahrimanes will be a very grim one.

"UFO technology portends the kind of confrontation that lies ahead. Man faces a bewildering armory of advanced technical devices, transcendental ability, and mind-bending powers. Armed only with mechanistic thought and an unbalanced technology, minus access to the ethers, the human posture for meeting this stupendous and unavoidable event is both unstable and inadequate."

Constable agrees with the many other serious scholars who have taken the time to assess the importance of these latter days: all depends upon expansion of human awareness.

"If man can be shown where the battlefield is, the nature of the terrain, and the ways in which he is already being assaulted in this inner war, then the right tactics and strategy can be brought to bear against the inimical forces.

"The objective of the Ahrimanic powers is to pull down all humanity. To this end, whatever political systems and philosophies exist among men are simply manipulated. Mundane party politics, political systems, and differences are therefore excluded from what is now being pointed out, because nothing definitive concerning the spiritual destiny of humanity can arise out of political studies or bickering."

It is Constable's contention that humans are constantly being seduced into doing the work of the nether forces because they simply do not know that such forces exist, let alone how they work *into* and *upon* Earth life. "Ignorance of this kind persists in the face even of the tumultuous events of the twentieth century and admits obvious motion of the human race toward a climactic period," Constable says. "Avarice is the main agency by which humans are seduced away from life, from positive work and positive thought. Incomprehension of spiritual forces and the institutionalized denigration of the spirit in formal education make humanity pitifully vulnerable to dehumanizing, life-negative, and destructive trends."

The UFO beings are, in Constable's view, etheric, rather than physical, beings. The struggle, as he sees it, is for the soul, rather than the planet, of man. "Entities riding etherically propelled vehicles and obviously in mastery of psychic control in all its forms devise contact encounters with ingenuous human beings who can be used in various ways to serve certain ends. Most of these encounters produce only chatter and bewilderment. These entities often hypnotize their victims, use them for various experiments, and then give posthypnotic suggestions of various kinds that will serve whatever end is sought.

"What is the overall consequence of these depraved encounters where an innocent human being is set upon by these weird humanoids? The world is led to believe that material craft are involved, if convinced at all. And if not convinced, then the contactee is another 'flying saucer nut.' Either way, the world gets a lie overlaid with confusion and ridicule while the humanoids depart from view . . ."

Constable warns us that the Ahrimanic emissaries appear everywhere "unrecognized and often aided by humans who don't know that the Devil is indeed alive and well—and coming to Earth within the lifetime of millions now living."

Constable is convinced that these Ahrimanic messengers inject themselves into UFO contact encounters in order to sow confusion and disorientation and to split and shatter those groups of serious-minded investigators.

"The battle for the Earth is a spiritual struggle but involves mighty temporal events," Constable writes in his book *The Cosmic Pulse of Life*:

> In this battle we are called to redeem the Earth
> through the creation of a new humanity—a humanity
> that will arise if we can restrain ourselves from shaping
> the new humans to the pattern of the past. Only with
> such a new humanity can a world order based on true
> brotherly love come into being and survive. By virtue
> of its armored-mask character, present-day humanity
> cannot sustain such a world order, but can only move a
> little in the right direction.

It would seem from the experiences of certain abductees that the citizens of Earth are encountering at least two representatives of extraterrestrial or multidimensional worlds. There appear to be those UFO intelligences who are genuinely concerned about our welfare and our spiritual and physical evolution, and there are those who seem largely indifferent to our personal needs and our species' longevity. Unless we are somehow perceiving different aspects of the same entities, then it might well be that our beautiful green oasis in space could serve as the prize in a war between worlds. If the Forces of Light and Darkness are about to square off on our turf, humankind could find itself the unwilling pawn in the ultimate battle—the final struggle between Good and Evil.

TO OUR READERS

MUFON BOOKS

The mission of MUFON BOOKS, an imprint of Red Wheel Weiser, is to publish reasoned and credible thought by recognized authorities; authorities who specialize in exploring the outer limits of the universe and the possibilities of life beyond our planet.

ABOUT MUFON

The Mutual UFO Network (*www.MUFON.com*) was formed by concerned scientists and academic researchers in 1969 for the specific purpose of applying scientific methods to the serious study of UFO sightings and reported human/alien interactions. MUFON's mission is "The Scientific Study of UFOs for the Benefit of Humanity" with the intent to unveil and disclose credible information free of distortion, censorship, and lies, and prepare the public for possible implications.

ABOUT RED WHEEL/WEISER

Red Wheel/Weiser (*www.redwheelweiser.com*) specializes in "Books to Live By" for seekers, believers, and practitioners. We publish in the areas of lifestyle, body/mind/spirit, and alternative thought across our imprints, including Weiser Books, and Career Press.